SpringerBriefs in Education

More information about this series at http://www.springer.com/series/8914

Jesse Bazzul

Ethics and Science Education: How Subjectivity Matters

 Springer

Jesse Bazzul
Faculty of Education
University of Regina
Regina
Canada

ISSN 2211-1921   ISSN 2211-193X   (electronic)
SpringerBriefs in Education
ISBN 978-3-319-39130-4   ISBN 978-3-319-39132-8   (eBook)
DOI 10.1007/978-3-319-39132-8

Library of Congress Control Number: 2016939570

Printed on acid-free paper

This Springer imprint is published by Springer Nature
The registered company is Springer International Publishing AG Switzerland

To,
Kate, Joseph, and Luca
What should I be doing again?

The forests of Massachusetts
For overgrowing them and setting us
on a path

The selfless friends
Foot masters and those who believe

Contents

Chapter 1
Science Education as a Site of Struggle

There are times in life when the question of knowing if one can think differently than one thinks, and perceive differently than one sees, is absolutely necessary if one is to go on looking and reflecting at all.

—Foucault (1985, p. 8)

Abstract In this introductory chapter I outline the purpose of this research brief, which is to think more critically about ethical engagement from a political, structural/poststructural perspective. I argue that science education must be seen as a site of struggle if 'wicked' twenty-first century problems are going to be engaged through/by education.

Keywords Science education · Politics · Ethics · Subjectivity · Power · Knowledge

Introduction

I struggled with the core idea of this book. What would be an original contribution concerning the topic of ethics to the fields of education, science education, and social theory? At first, I wanted to develop a critical, activist framework for ethical engagement in science education. One that would cast the self against the political backdrop of identity, politics, and power when making decisions about our collective wellbeing. My motivation exceeded the growing demand for science and technology to engage with ethics. Societies in the twenty-first century face new ethical challenges related to impending environmental catastrophes, growing social inequality, and biotechnologies and biomedical research (Žižek 2011; Rose 2007). These problems persist partly because many people still cast reality on one side (economics, science, mathematics, human nature) and idealic dreams on the other (philosophy, art, morality, and politics).

The best interests of the planet, however, involve mixing 'dreams' with 'reality'; or realizing the two, the virtual and the actual, are always linked together. But an

J. Bazzul, *Ethics and Science Education: How Subjectivity Matters*, SpringerBriefs in Education, DOI 10.1007/978-3-319-39132-8_1

engagement with ethics in/through education is not at all new, as one could say that the fields of critical pedagogy, civic education, science and society (SSI, STSE), policy studies, bioethics, and special education, etc., are all fundamentally engagements with ethical ways of being. Indeed, the educative act is all about having/communicating/developing an *ethos*. No, it would be impossible to lay out a new ethics for (science) education here: which is why I rejected Springer's™ suggestion for the title of the brief: *Ethics for 21st century biology education*. Such a volume would have to be written by many scholars, perhaps consisting of a collection of the ethical issues related to science at our current socio-historical location. It is also likely not useful to re-engage debates about the epistemological status of scientific knowledge and its relation to reality, what in the 1990s were called the 'science wars'. This debate seems to have fizzled, likely because there is more awareness of the pressing social and environmental issues at hand. Critical questions of science education in past decades often focused on the limits, exclusionary potential, or underlying biases of the objective sciences. Instead, I would like to argue that the important field of science education *is*, and will increasingly be, a *site for ethico-political engagement* with the major problems of our century: environmental destruction and growing social inequality. Let's take a paragraph or two and unpack this statement.

Science education is a *site* of struggle because of its connection to power exercised in the form of objectivity legitimated both by the governmental apparatus of educational institutions and the 'truth value' of scientific knowledge. Science education therefore is part of an apparatus of governance where what is considered objective merges with what governments think is useful for populations to know. In other words, science education very much involves the politics of knowledge. On the terrain of social and environmental justice, science education should be seen as part of a constellation of sites for critical engagement along with the street, the organization, community, park, social media, ballot box, mainstream media, novel, gallery, labour union, farm, etc. These sites can all be tied together, and to use the activist parlance of our times, they can all be occupied. Due to the major problems of the twenty-first century, and the new ethical demands that now face societies, it is important for educators to understand how students are positioned as ethical actors and decision-makers in/through science education.

By 'ethico-political', I mean to include, alongside ethics, the political battles surrounding social struggles of marginalized, oppressed, exploited humans and animals, etc. Not just so that science educators and students engage political issues, but for educational communities to rethink what it means to be political in/through science education. To make the sociopolitical *the* stakes of science teaching and learning. Science Education as a site of both control and resistance means there must simultaneously be a *politicization* of science education. And while it is recognized that science and education is never politically neutral (Darder 2014), what is becoming more apparent is the need to turn our planet around from the course of climate change and the well documented changes that mark the *Anthropocene* (Lewis and Maslin 2015). There is now an ethical warrant to employ science and technology education against forces that further the destruction of the planet and widen social inequality.

This brief will explore the dimensions of this 'ethical subject' largely employing the work of historian and philosopher Michel Foucault. I mean ethical subject here in two ways: first in terms of the topic of ethics as it relates to science education; second, in terms of subjectivity, more specifically a subject's (individual's) relation to self, others, and the world. Specifically, I argue that these 'relations of self' must be conceived of in ways that are democratic, collectivist, and egalitarian. In short, ethics in science education must interrogate not just ethical codes, but the dimensions by which subjects are brought to operationalize ethics in issues related to science. In this research brief I argue that, for the sake of the planet, these ethical dimensions of self must take a more politicized form.

I will address the topic of ethics in Chap. 3, not by beginning with a theory of ethics or what ethics should be, but by discussing how the discourses of Canadian biology textbooks can work to constitute particular outlooks related to science and ethical issues (see Bazzul 2014). According to the study, which I conducted as part of my doctoral work, the range of potential ethical actions a student might take, as outlined in discourse, is a limited subset of possible forms of engagement (one overall point being that the ethical choices afforded to subjects are always limited). In tracing what possibilities are open and closed through the discourses of biology textbooks, we can ask after the limits of ethical engagement for the purposes of seeing what else is possible. How did it come to be that students deal with one ethical issue and not another? Why do students engage these issues in some ways and not others? Who are students expected 'to be' before they engage ethical issues in science education? The larger goal of analysis discussed in Chap. 3 is to ask/develop questions of policy and curricula that outline how students are to engage ethically with social and environmental issues related to science. However, in addition to the question of ethics, it is important to think a little about the political orientation, as well as some practical features, of ethics related to science.

An important practical, everyday feature of how someone engages ethics is *subjectivity*. We can begin to think about subjectivity by simply considering the beliefs, values, attitudes, and outlooks that a person holds. Educational institutions are central in the reproduction of subjectivities, and are what philosopher Althusser (1998) calls an Ideological State Apparatus. Educational institutions, discourses and practices play an integral part in determining how generations of peoples come to see aspects of their identity (race, sex/gender and sexuality, spiritual beliefs) as salient, as well as legitimating political orientations and ethical actions. Education, including science education, is therefore a site of engagement for the constitution, and reworking, of subjectivities. To contribute to a more just world, education must constitute subjectivities that are altruistic, caring, and communal. Scholars in education should pay specific attention to how subjectivities are constituted through discourses and practices of education. This cannot be overstated. While many educators realize that schooling shapes the hearts and minds of future citizens, scientists, etc., there has arguably been insufficient pushback to the erosion of public education, racial segregation in schools, and lack of environmental education (Lipman 2013;

Greenwood 2008). To put it another way, schooling has not sufficiently engaged these issues at the level needed for our long-term survival.

I establish a basis for understanding subjectivity from a critical, structuralist/ poststructuralist perspective,[1] along with a general methodology for analyzing discourses of science textbooks, in Chap. 2. I use broad categories of politics and identity to discuss subjectivity, categories that should only be thought of as abstractions, such as sex/gender and sexuality, political views, race, and geopolitical positionality (e.g. Global North or Global South). Subjectivities are constituted along many dimensions affectively, structurally, and materially. In Chap. 3, I revisit this methodology, and the textbook study where it was developed, to outline how discourses of science/biology education can work to constitute an ethical subject or subjectivity. In this brief, I will specifically discuss the role of discourse, and the practices discourses outline, within institutionalized science education. To do this, I draw heavily from the work of Foucault because his thought helps us understand social institutions, power, and subjectification (the process by which individuals are constituted as subjects). Foucault may also be overlooked as a philosopher of science as his work helps trace the relationship between power, governance, and knowledge. This includes scientific knowledge and how it brings people to think about themselves, and conduct their lives, as living, biological beings through discourses of biology, ecology, genetics, health, medicine—irrespective of whether this knowledge is true or not (a point I make in this chapter). All in all, I take a specific approach to the problem of science and governance through ethics, politics, and subjectivity.

Chapter 4 lays out a sociopolitical context for the constitution of subjectivity through schooling. Why are some subjectivities constituted and not others? What are the broader contexts that govern schooling? To address these questions I employ Foucault's concept of *biopolitics*, which essentially means looking at how modern governance revolves around the control, and resistance, of particular forms of human social life and being. Institutions of schooling employ particular rationalities, structures, discourses, and practices to 'conduct conduct', including modes of self-governance that form part of the basis Foucault would call *governmentality* (Foucault 1982; Foucault and Senellart 2010). What should be of interest to science educators is that technologies and strategies of modern governance use discourses of science to forward particular aims, for example the notion that humans are a biological species. Science education for the public introduces many pertinent social, ethical, and political topics to students including population control, agricultural policy, hygiene, appropriate sexual mores, and the genetics of 'relevant' human behaviour. The power exercised through these discourses can be thought of as *biopower*, how life itself becomes the target of governance. Consequently, to understand how science is entangled in relations of power in modernity, and to engage with these relations of power, a biopolitical approach is necessary. Biopolitics is not a political

---

[1]I take Claude Levi-Strauss' (2013) view of structuralism, namely that structuralism endeavors to find something repeatable or regular about phenomena, language, institutions. In this sense theorists like Althusser, Foucault, and even Butler operate as 'structuralists', even though their thought escapes the boundaries of this school of thought.

stance, but a set of theories that can help make sense of how power and governance are exercised through science education.

Part of what drives this book is the belief that science education does not place enough value on the social, historical, cultural, spiritual, philosophical, and political backdrop that animates, drives, and pervades science—particularly when it is under the control of stifling standards-based policy. If science education is a site for ethico-political engagement, we need to engage a diverse array of philosophical traditions, from indigenous thought to feminist pedagogies. This brief could be said to be an interdisciplinary exercise. Like other justice oriented, critical work in science education, it does not seek its warrant from concerns already present in the literature of science education, the policy documents of major organizations that deal with science education like the NRC in the United States, or the opinions of senior scholars in the field. This brief is an attempt to address some underlying assumptions about ethics and politics in science education in order to bring change, albeit small, to science curriculum and pedagogy.

This brief challenges the idea that science education is just a place where kids learn 'cool things' about how the natural world works and acquire the skills they need to get a good job. Of course, science education does these things! But much more is happening. Ethical outlooks, attitudes and beliefs—subjectivities—'take shape' in science classrooms; and these classrooms may very well play a major role in whether or not local and global communities are able to engage ethical issues related to science. As I shall discuss in the Chap. 3, remaking science education a site of ethico-political engagement requires that we continually look to new social, ethical, and political forms of engagement. In Chap. 5, I argue that educators need to politicize ethical engagement if science education is to play a key role in shaping a future that engages ethically with the sociopolitical issues of our time.

I have tried to write this brief in a way suggested by my colleague Alex Means: 'out of my head' (personal communication, June 15th 2015). This could mean a lot of things! And I will say that all interdisciplinary projects are very difficult—in this case combining science education, critical theory and ethics. Ideas and methods that transect disciplinary boundaries will inevitably seem out of place, strange, irrelevant, and a misfit for many quite used to the discipline of science education. Interdisciplinary scholars know how hard it is to come under fire from two or more sides. We take a piece of work to one 'camp', and they tell us it belongs somewhere else… we go there only to be told to turn back the way we came. We take the risk of misrecognition as we cross borders. So, if you were to say that the justice-oriented political agenda, relatively novel ideas about ethics, and the social theories found in the following chapters do not belong within the discipline of science education, I will have, in part, achieved my goal.

References

Althusser, L. (1998). Ideology and ideological state apparatuses. In J. Rivkin, & M. Ryan (Eds.), *Literary theory, an anthology* (pp. 294–304). Malden: Blackwell.

Bazzul, J. (2014). Tracing "ethical subjectivities" in science education: How biology textbooks can frame ethico-political choices for students. *Research in Science Education, 45*(1), 23–40.

Darder, A. (2014). *Freire and education.* Chicago: Routledge.

Foucault, M. (1982). The subject and power. *Critical Inquiry, 8*(4), 777–795.

Foucault, M. (1985). *The use of pleasure.* New York: Pantheon Books.

Foucault, M., & Senellart, M. (2010). *The birth of biopolitics: Lectures at the College de France, 1978–1979.* New York: Picador.

Greenwood, D. A. (2008). A critical pedagogy of place: from gridlock to parallax 1. *Environmental Education Research, 14*(3), 336–348.

Lévi-Strauss, C. (2013). *Myth and meaning.* New York: Routledge.

Lewis, S., Maslin, M. (2015). Defining the anthropocene. *Nature, 519,* 171–180. doi:10.1038/nature14258.

Lipman, P. (2013). *The new political economy of urban education: Neoliberalism, race, and the right to the city.* Chicago: Taylor & Francis.

Rose, N. S. (2007). *Politics of life itself: Biomedicine, power, and subjectivity in the twenty-first century.* Princeton: Princeton University Press.

Žižek, S. (2011). *Living in the end times* (Rev. pbk. ed.). London: Verso.

Chapter 2
The Constitution of Subjectivities: Discourse, Practices, and Repetition

> *…[o]ne of the places most likely to provoke a questioning of the scientific landscape is that of the examination of the subject of science and its psychic and sexed implication in discourse, discoveries and their organization.*
>
> —Luce Irigaray (Irigaray and Bové 1987, p. 79)

Abstract In this chapter I outline a theoretical approach to subjectivity as it is constituted through power-infused discourses. Discourse constrains and affords certain choices for individuals and therefore works to allow certain outlooks, beliefs, practices, and ways of being (and not others). I argue that how educational discourses work to constitute subjectivities should be a focus of educators and researchers if they hope to gear education toward the goals of justice. I discuss how a focus on subjectivity can be useful in determining what identities are constituted and valued in science education.

Keywords Subjectivity · Discourse · Foucault · Science education · Power · Knowledge

As mentioned in the introduction, subjectivity is a broad term that refers to beliefs, attitudes, outlooks, convictions, subconscious tendencies, orientations, proclivities, understandings, etc., that people may hold (Hall 2004). Although I will articulate a particular view of subjectivity here, it is very important to maintain subjectivity, its construction, importance, and 'character' as a site of contestation (Butler 1995). There is no 'correct' way to look at subjectivity and/or the self, and there is also no subjective category that has an absolute ontological or epistemological foundation in reality (though there may be some that are more useful). That is, although we can think critically about a 'political subject', a 'neoliberal subject', or a 'subject of sexuality', these categories are organizing abstractions that allow us to understand subjectivities as they are constituted by the social order. This point is subtle, and perhaps obvious, but is important to mention because it underlines one key aspect

© The Author(s) 2016
J. Bazzul, *Ethics and Science Education: How Subjectivity Matters*,
SpringerBriefs in Education, DOI 10.1007/978-3-319-39132-8_2

of subjectivity that should be kept in mind: subjectivities are ever-changing, hybrid, and never totally constituted with no hope of thinking or acting differently.

Following Foucault (1972, 1982) and Butler (1997), I argue that attention to how particular kinds of subjectivities are constituted through discourse and practices is a useful form of analysis for those interested in how oppressive and exploitive social institutions, networks, and private interests work to produce the very kind of 'being' required to maintain a particular social order. In this chapter I focus on how subjectivity is constituted through limitations and affordances, how these limitations and affordances are delineated by discourses and the repetition of practices, which are often outlined in discourses. This is not to say that considering subjectivity from the perspective of phenomenology, freedom and becoming, or more cognitive perspectives is not just as important. Certainly, many assumptions and conclusions I will make are quite compatible with other ways of viewing subjectivity.

I slowly became interested in subjectivity because of its centrality to struggles over equality, environmental justice, and being different. Thinking about 'the subject' has been central to social theory since the time of Karl Marx. Understanding that subjectivity is constituted exceeds the notion that identity is socially constructed, or that human beings are socialized into particular roles, norms, and positions. Although these things are more or less true, the fine, but crucial, distinction that must be underlined is the difference between *individual* and *subject*. The work of Althusser (1998), a close mentor and friend of Michel Foucault, is useful in explaining how individuals are turned into subjects through ideologies manifested in apparatuses and institutions of power (school, hospital, factory/workplace, church, family, barracks, etc.).

I think it important to spend a little bit of time developing how individuals are constituted as subjects because, although subjectivity makes an appearance in the science education literature, it is either not discussed in-depth or is approached from different perspectives, such as hybridity, worldview, or 'as a phenomenon'. I find these other perspectives very useful, though they have their difficulties. Hybridity, the fact that each subject is simultaneously multiple, is part of the reality of subjectivity in postmodernity, but recognizing hybridity may also be just a way of recognizing the logic of late capitalism and the cascade of various 'identities' propagated by the market (Jameson 1991). Unless hybridity is politicized to promote multiple, 'other' becomings, or political recognition, it can remain a banal concept for educators working towards justice (and the same goes for subjectivity in general)! The literature on worldviews (see Matthews 2009) offers us a useful construct; however, in science education this literature primarily appears to only question whether a 'scientific' worldview is itself a cultural or situated worldview. Differentiating between 'scientific' and 'other' worldviews may also work to detach sociocultural considerations and political dimensions from scientific worldviews. Lastly, a phenomenological perspective can root students, subjectively speaking, in their 'lifeworlds' and this makes phenomenology an important tool for educators (Bazzul 2014c; Roth 2014). However, our own view of our lifeworlds is always already constituted, such that the 'subject' of a lifeworld is always an effect of this lifeworld. This is exactly Althusser's central point concerning subjectivity: we cannot speak outside of ideology. We are always already ideologically constituted as subjects through ideological apparatuses: the institution of schooling being one such apparatus.

The Importance of Althusser, Butler, and Foucault's Work

Althusser is an important theorist because his work arguably helped form the basis of the work of Michel Foucault, Judith Butler and subsequent critical scholars. For Althusser, ideological apparatuses (e.g. factory and school) constitute the subject through mastery of particular practices. The ideology of apparatuses is not to be thought of as 'false consciousness', but the social fabric of reality, perpetuated through practices (Althusser 1998). These practices must be repeated, and through this repetition the subject is continually (re)produced (Butler 1997). The strength of this point can be seen in the old maxim: *Your actions 'speak' louder than words*. In an ideological apparatus such as a school, it is what students, administrators, teachers, and workers do that constitutes what they 'believe'. Althusser uses the example of religious practices to make his point. He claims it is not so much that a subject, say a child growing up in the Catholic Church, comes to believe in the body of Christ before first communion, but rather through repetitive kneeling, praying, recitation. This is to say that through ideology, rendered through discourses, a subject comes to acquire a certain set of practices, also rendered through discourses, which retroactively confirm beliefs. In light of this, it is pertinent to ask what kind of repetitive work science students do in classrooms, laboratories and inquiry settings. Elsewhere, I explore this question in relation to how current forms of science education may work to constitute a 'subject of (science) labour'; someone who sees work in science as a personal investment and disconnected from community concerns (Bazzul in press).

To summarize, it is ideology, as rendered in discourses and transmitted through practices in an apparatus, that constitutes individuals as subjects. According to Althusser, this happens through a process of *interpellation*, which is essentially the continual positioning of a subject (read student) in the social order. Althusser refers to this process of continual positioning as hailing, where subjection to the ruling social order, ideologies, or 'law' is confirmed. Althusser uses the example of the police hailing someone, shouting: 'hey you there!' As soon as the subject *turns around* they confirm their status as a subject of the law—the law being understood as the authority/legitimacy of the current social order. In schools, hailing occurs even in the very act of taking up the position of student. The question is not necessarily whether hailing is 'good' or 'bad', indeed most of us answer the 'hail' of our own names (for the most part given and not chosen). For good of for bad, interpellation can be seen as a reality of subjectification.

Since we all of us are interpellated in institutions, social groups and organizations, before we come to realize it, we can think of subjects as *always already* constituted. In educational institutions students will be constituted as subjects in a multitude of ways at any given time. Furthermore, subjects attach meaning to, and find identity in, subjection through the family, nationality, cultural background, disciplinary regimes, social organizations (clubs), sex/gender norms, etc. Subjection is therefore a process that controls modes of being for the subject, yet also comes to be the very conditions for the subject's freedom: "I am this; My true self belongs here". Any critical analysis of subjectification involves a process of *asking after* how it comes to be that we find certain practices, ways of speaking and being, as normal, acceptable, and thinkable—and not others. Given that we can only be critical of our subjectivity after it has been constituted, the goal of a critical scholar is to render what seems commonsensical, strange (Foucault 2003b; Butler 2002). To become different means to intervene in subjectification practices, changing them both at the site of the self and the world. This is no easy task, as identifying and changing practices by which one comes to find their own identity is much more difficult than it may appear.

Butler (1997) elaborates a few ways through this 'problem' of subjectification by identifying at least three ways where change is possible. First, different discourses can come together and produce unintended or unexpected results. This is one way to look at Fausto-Sterling's (2012) work on sex/gender, sexuality, and developmental biology. Fausto-Sterling is a biologist who adopts a critical sociological lens to help develop new biological understandings of how people come to gender/sex identities. This is a significant move away from sex/gender studies in biology that often retain a male, western, and heterosexist gaze (Fox-Keller 1996). Discourses themselves can sometimes have unintended consequences. It is useful here to think of Foucault's (1981) example of how the 'homosexual', as a category of reality, emerged in the 19th century. Through discourses of hygiene, moral conduct, racialization, colonization, nationalism and biology, 'homosexuality' came into being precisely when the intention was to police it. Consequently, new cultures, practices, ways of being and forms of resistance developed around this social category. This is to say that discourses can sometimes exceed their normalizing aim.

The second and third ways in which subjectification can be disrupted or changed occur somewhat in tandem. Since there is a tendency for a subject to be hailed, this hailing can be disrupted. Furthermore, subjectification requires the repetition and re-iteration of discourses and practices. In order for a subject to continually recognize themselves as a subject there must be a continuous iteration of what allows the subject to be recognized as that subject. For example, if Simons (2006) is correct, schools are now working to produce an entrepreneurial subject through self-investment and the stripping away of public investment. We can expect aspects of this subjectification process to occur relatively frequently in multiple places, e.g. testing justification, career discourses, course choices, assessment practices, etc. The tendency for a subject to turn toward the 'law' of the social order, combined with the fact that discourses and practices must be repeated for subjectification to be maintained, means that the site of interpellation, hailing, and their repetition is simultaneously the site of resistance and change. For example, sex/gender and sexuality is something that must be performed, according to Butler (1990), in iterative fashion. Since there is a limited number of iterative moves available in this performance, innovation in practices, identification, love, etc., happens at sites already designated for this performance (e.g. mate choice, dress, body movements, affect(ions), speech). And although any one 'move' might seem banal at first, when synthesized and collected over time they collectively usher in new ways of being. This is why it is vital for educators to understand and intervene in subjectification practices—so as to create opportunities for students to become different.

Foucault extended Althusser's concepts in a number of ways, backing them up with extensive historical analysis (Foucault 1972, 1977). For Foucault, an apparatus is more than the usual institutional 'site' of subjectification and control—the family, school, hospital, factory, office, bureau, barracks, place of worship, etc. An apparatus is a confluence of technologies of power, discourses, practices, and material arrangements that constrain and direct what is possible—'truth' discourses, or objective knowledge, help circulate the effects of power in an apparatus because they outline the limits of 'the possible' through affordances and silences. The more 'objective' or true a discourse is, the more it can be employed to distribute the effects of power through a discursive field that institutionally recognizes some statements as 'in the true' and other statements as not 'in the true'. A statement, in the Foucauldian sense, is any utterance with institutional force, that is, backed up by a discursive field and institutional arrangements that classify this statement as 'in the true' (Foucault 1972). Thus, objective statements, whether they are in fact 'true' or not, require a specific social, cultural, and political discursive field to qualify them as such. Truth statements support power relations by giving legitimacy to policy, government structures, lifestyle choices, and the organization of social life by those in a position to employ this power (Foucault 2003a). Power does not reside with any one person but is distributed in a field of relations.

Discursive statements circulate in social communities outlining a particular field of objects, subjectivities, and what can be operationalized and acted upon. While some discourses are very specific to a locality, statements from discourses are generally present and at play for members of a community.[1] This has implications for the status of an author or speaker. If discourses circulate among communities, then statements are ostensibly available for use. This moves the responsibility for discourse away from individuals and authors as 'originators' of discourse (e.g. 'independent genius or insight') toward the idea that texts and speech are like *captures* of discourse circulating in a particular social, cultural space. This view of discourse does not totally oppose the idea that different authors produce unique texts, but merely diminishes how unique the thoughts of one individual can be when most of what is thinkable or utterable circulates in a community.[2] If thoughts, ideas, attitudes, beliefs circulate as discourse, then subjectivity can be thought of less as a product of a unique mind free to think anything it wants, but rather as something constituted in particular ways by apparatuses in the social order. An apparatus can also be thought of as all of the ways, both micro and macro, that distribute the effects

---

[1]This makes Foucault's notion of discourse more plastic that Althusser's notion of ideology (see Mills 1997).

[2]And it is often the case that at the time someone comes up with an idea, many others are thinking along the same lines with the concepts, data, and procedures available!

of power. This 'innovation' is useful in social analyses because the way power is exercised in society can take more forms than a 'superstructure' or an ideology that must always have, at its root, an economic imperative. Power, in the Foucauldian sense, is not something some people or offices *inherently* hold and others do not; rather, it can be thought of as a relation that people and institutions hold to others and the world. Power governs an entire field of relations across a social 'body', from aspects of identity (race, sex/gender, sexuality, etc.) through normalizing discourses, to how we see ourselves as subjects in ethical and political contexts (labour, self actualization, life purpose, etc.). Hopefully we can begin to see a key relationship in Foucault's work between *discourse*, *truth* (objectivity), *governance* and *power*, and the constitution of subjectivity—and how an ethically and socio-politically engaged science education would ask after how subjects are constituted through governing apparatuses of power.

Consequently, we can ask: What discourses are circulating through state-corporate apparatuses of education? How do these discourses afford some possibilities and not others? How do education communities think differently? This summary is only a brief introduction to the constitution of subjectivity through discourse, power, and governance from a structuralist/poststructuralist perspective. I will now outline a textbook study that examined how discourses of textbooks could work to constrain thought and action. This section is important, as it will offer a context for the way in which I engage the topic of ethics in Chap. 3. To conclude this chapter I will return to the topic of subjectivity as a central concept for engaging issues of justice in science education.

Subjectivity and the Discourses of Biology Textbooks

During my dissertation work at the University of Toronto (OISE), I set out to see how discourses of four Ontario secondary school biology textbooks worked to delimit thought and action for students and teachers. This interest stemmed from my time as an undergraduate student of biology and my work as a science teacher, where I became uneasy with certain discourses and practices. It was then that I first noticed that textbooks, and science professors, would often touch on real world issues to explain or contextualize concepts. What was noticeable after a period of time was that these real world issues were being discussed with(in) a discourse of strong objectivity. So while political, ethical, and cultural phenomena are multifaceted, socially constructed, and culturally situated, they are spoken about with a discourse that connotes objectivity, *'this is how the world is'*. Unlike other discourses, scientific discourses, especially the sanctioned knowledge of textbooks, often operates as if it were value-free. I became concerned with the oppressive nature of this confluence of social issues and objectivity as it relates to the marginalization and exploitation related to race, class, sex/gender, sexuality, geopolitical status, ability, etc. Some publicly known examples of this can be seen in the comments of Harvard University president, Lawrence Summers, declaring women academically 'less able', or grosser still, James Watson's comments about African intelligence (it should not be overlooked that they were both made by white, Anglo-American males). In both cases, the speakers referenced aspects of biology to reference their oppressive claims, and did so with the matter-of-factness fitting of a racist or sexist; or someone who thought they had 'truth' on their side.

My critique of science education discourses is centered in biology because I have studied the discipline, but also because biology is a science more connected to apparatuses of governance and power than other natural sciences. Although all science operates within a social, cultural, and political frame, the ability of some sciences to distribute the effects of power may be higher than others because they more often deal with human concerns (of course this distinction is slowly breaking down with the increasing recognition that non-humans have the right to flourish). In short, biology is where scientific knowledge more often meets the social world and therefore is a site where relations of power can be discerned and reworked. In my analysis of textbooks, I took a Foucauldian approach to discourse analysis. I will now outline this methodology in order to show how I came to some of the results presented in Chap. 3, and provide one way a researcher could approach texts/discourses in a critical way.

Methodology for Analyzing Science Curriculum/Textbooks

I want to outline some basic points about Foucauldian discourse analysis because they describe aspects of discourse and subjectivity relevant for other social analysis, and the results related to ethics in Chap. 3. The methodology for this study is described in more detail in an article entitled "Critical Discourse Analysis and Science Education Texts: Employing Foucauldian Notions of Discourse and Subjectivity" (Bazzul 2014a, b, c). The texts in this analysis were four secondary biology textbooks approved by the Ontario Ministry of Education: *Nelson Biology 11 College Preparation* (DiGiuseppe 2004); *Nelson Biology 12* (DiGiuseppe 2003); *McGraw-Hill Ryerson Biology 11* (Dunlop 2010); *McGraw-Hill Ryerson Biology 12* (Blake 2011) (Ontario Ministry of Education 2008).

Textbooks are important locations for analysing the cultures of science, because as the famous historian of science Kuhn (1996) notes, textbooks are the sites "where each new generation of citizens and scientists learn to practice their trade" (p. 1). They should therefore be seen as 'open' texts (Eco 1989). That is, open to multiple interpretations and meanings. For example, Ninnes' (2002) analysis of space science and national politics in science textbooks demonstrates how textbooks discursively limit the range of views students can legitimately hold. Texts, even if seen as banal representations, have social, political, and material effects in the world. Though these effects may not manifest within every individual or every situation, they still hold the overall potential to be causal to actions and ways of thinking (Fairclough 2003). Science education is infused with similar discourses that exist within other sites of subjectification such as other forms of education, media discourses, etc. They are one site amongst many sites of social and political (re)production and engagement. The point is that science education is not an exception, and should not be exempt from analyses that ask about how discourses constitute the outlooks of students. On the contrary, due to the ever-growing political interest in science and technology, science education discourses will become increasingly important in shaping future generations. Lather (2012) sums up research that asks after how discourses in textbooks constitute subjectivity with the provocative question: *Who does the textbook think you are?*

As mentioned above, a statement's institutional force is tied to its level of objectivity. Science textbook discourses are therefore somewhat unique in that they carry a 'double-sanction' from government (approval) and science ('objective knowledge'). For Foucault (1972), discourse outlines the set of possible actions for a subject, thus one form of analysis would focus on delineating the field of choices open to a subject, as well as where and how those choices are dispersed (p. 40). Power is exercised through education and science education discourses, not as a coercive force, but as a field of 'positive', commonsensical possibilities. One of the major innovations Foucault (1980a, b) made in the field of social theory was to recognize that power is productive (as well as repressive); otherwise, people would not "manage to obey it" (p. 36). And although power is exercised 'positively' in the social world, there always exists *the potential* to rework relations of power and the possibilities opened to subjects.

A Foucauldian archaeological approach to reading texts is specific in its adherence to several general guidelines (Kendall and Wickam 1999; Foucault 1972; Bazzul 2014a). I shall outline four of these guidelines here because they are crucial to approaching science texts, subjectivity, and discourse from a Foucauldian perspective.

Minimizing the Author Function: A Foucauldian approach does not attempt to intuit the intentions of writers, because Foucault's notion of discourse implies that texts are not unique creations of individual minds—an assumption of traditional, modern, western views of texts (Foucault 1984). Organizing texts by authors and publishers can become arbitrary, as subjects always speak within larger discourses. Dropping the author function allows researchers to assume discourses found within texts exist at a larger level, thereby allowing them to analyze statements across a wider variety of texts belonging to similar regimes of power/knowledge. The analyst suspends the idea that the goal is to try and find the real intention of the author.[3]

Reading the Surface of the Text: Instead of focusing on intentions and 'deeper meanings', analysts focus on what the surface of the text says literally. Author intentions and an assumed unity to a text can distract the analyst and obscure the process of finding the contours of a discourse. As Kendall and Wickham (1999) maintain, a Foucauldian analysis "cannot go beyond this discursive 'surface' to a 'deeper inside'

---

[3]Again, this is not to say that the author function does not play a role in textual analyses, nor that the identity of an author is unimportant to analysis of curriculum and policy. For example, when no specific author is given, textbooks take on more of an authoritative quality; their author becomes less of a "who" than a "what."

of 'thought': the surface is all there is" (p. 37). Archaeology remains at the level of discourse.

Archaeology as Cross-section: An archaeological analysis is a cross-section of a particular discursive regime, or on a smaller-scale, discursive formation—in contrast to genealogy, which looks at changes in regimes and their apparatuses and discourses over time (Foucault 1980b, p. 85). The goal is not to discern whether discourses or statements are true or false, but rather to examine how they constitute individuals as subjects. Truth is important insofar as it outlines "how what is said to be true and false makes things ordered and pertinent" (Foucault 2003c, p. 252), that is, "the effects in the real world to which they are linked" (p. 257). Foucault's (1970) aim in archaeology was "to reveal a positive unconscious of knowledge: a level that eludes the consciousness of the scientists yet is a part of scientific discourse, instead of disputing its validity and seeking to diminish its scientific nature" (p. ix). Archaeology provides a sketch of how institutions exercise power and authority.

Statement Specificity and Relations Between Statements: In an archaeological approach, analysts isolate relations between discursive statements. In doing so, the notion that a text represents a commonsensical whole is suspended. The analyst focuses on the specificity of a statement's emergence in a text, the relation between the sayable and the visible, and the rules by which statements about the social world are formulated, and how statements delimit a particular field of objects, ways of knowing and doing, as well as norms of conduct, explanation, communication, etc.

Some Directions for Critical Analysis

When conducting Critical Discourse Analysis from a Foucauldian perspective, we want to find out what subjectivities are given importance in discourse (Connelly et al. 2008). To do this, flexible theoretical frameworks are needed because different aspects of subjectivities need to be understood from different perspectives. As Fairclough (2003) states, the analysis of texts "should be seen as an open process which can be enhanced through dialogue across disciplines and theories, rather than a coding in the terms of an autonomous analytical framework or grammar" (p.16).

The focus of this study looked at how these textbooks constitute subjectivities related to ethics (Bazzul 2014a, b, c), sex/gender and sexuality (Bazzul and Sykes 2011), neoliberalism (Bazzul 2012), and neocolonialisms. Again, these categories are abstractions that try to capture aspects of subjectivities—or in Patti Lather's terms, 'who these textbooks think we are' (Lather 2012). I will outline these categories below, along with some of the theoretical lenses used to make sense of these discourses.

Intersections of Sex/Gender and Sexuality and Race

An analysis of one of the textbooks, *Nelson Biology 12* (DiGiuseppe 2003), led to a kind of exposé of the way textbook discourses privilege heterosexuality and strict, almost mythic, sex/gender norms (Bazzul and Sykes 2011). Not only is heterosexuality the de facto sexuality, but all other traces of other sexual practices, gender/sex identities related to humans are silenced. Instead, idealic forms of 'male' and female are found throughout the text. Students using these texts are presented with social, biological, and sex/gendered worlds from a binaric perspective. These texts exercise power through authoritative, objective, and normative discourses in terms of what can be thought in relation to one's own conception of gender, sex, and sexuality (Foucault 1980a).

Later in the analysis, images in the textbooks were examined for representations of race and sex/gender. Although images appeared to represent diversity in terms of sex/gender and race, the frequency of images of people performing science work, as well as 'lay roles', were highly skewed toward 'white male' representations. When images of people of colour were coded into binaric 'male/female' categories (the irony of which was not lost), the frequency of images were much more skewed toward males in both lay and science work images (Bazzul 2013). This demonstrates that sex/gender operates differently in relation to race in the discourses of these biology texts if we accept that images are also texts. It is very difficult to categorize, and then comment on these images as this can work to reproduce racializations and (hetero)sexist, binaric categories even though the aim of the image analysis was to identify the presence of masculinities, whiteness and privilege. The results demonstrate clear hierarchies in how racialized, sexed/gendered groups are represented, with white males dominating these representations (Bazzul 2013). Educators should ask how students who don't identify with being white, male, gender-normative, or heterosexual are positioned in relation to science and science work. How do they come to see themselves in relation to objective, power-wielding discourses of science education? Feminist philosopher Luce Irigaray (Irigaray and Bové 1987) argues that science and philosophy begin with, and maintain, a Eurocentric male subject position (perhaps it is appropriate to add white, heterosexual, ableist, etc.). Do we as educators see Irigaray's argument reflected in the very discourses of science education?

Neocolonialisms: The Colonizer and the Colonized

A case could be made that these Ontario textbooks also work to normalize along Eurocentric, western hegemonic lines; for instance, through the delegitimization of local, 'traditional' knowledges. A key critical question for science education in a globalized world, could be how far it goes in establishing the supremacy of a Eurocentric, modern western view of the world over all others. Said (1978) implicates scientific discourses in the production of colonized and colonizer subjects (peoples under colonial rule) and objects (the casting of the world from the perspective of colonizers). Stereotypes fix relations for colonized subjects by connoting disorder and degeneracy, and like other forms of subjectification, must be repeated over and over again in various locations to 'stick' (Bhabha 1994). Homi Bhabha insists that both the colonized *and* the colonizer are constituted in colonizing discourses, as he maintains that it "is difficult to conceive of the process of subjectification as a placing within Orientalist or colonial discourse for the dominated subject without the dominant being strategically placed within it too" (p. 72).

A colonized and colonizer subjectivity may be constituted by these biology texts through a division of those who *can* know or make claims in science, e.g. the global north, and through the trivialization of indigenous knowledges. The latter is done by reducing these knowledges to what can only be 'legitimized' from the perspective of modern western science (Bhabha 1994). For example, an exercise concerning deforestation in *Nelson Biology 12* (DiGiuseppe 2003, p. 180) tacitly outlines two peoples in the debate. The first, scientifically minded, concerned citizens who are able to see the true nature of the situation and act accordingly, and the second, others who are incentive-driven and short sighted. In other words, the discourse of the text associates science and concerned citizenship on one side—and perspectives that only people of the Global south could have on the other—demonstrating that the contexts of science cannot be divorced from geopolitics. Another example from *Nelson Biology 12* devalues non-industrial, non-agricultural societies in statements such as this: "The geographic distribution of suitable wild plants and animals largely determined the regions in which agriculture would arise and where human societies would follow" (DiGiuseppe 2003, p. 700). What is the effect of such a discourse on the subjectivities of students who do not live in, or come from, agricultural or industrial societies? What is the effect on those who do live in these agricultural or industrial societies in terms of their outlook on others who live in different types of societies? Again, a big question in relation to colonialism that science educators should ask is whether science education is part of larger political processes of subjugation. Or in biopolitical terms, how modern governance, exercised through education, targets some in society for the overall sake of the 'body politique'. I will pick up this thread again in Chap. 4.

Constituting Depoliticized Neoliberal Subjectivities

Neoliberalism as a political project, ideology, and as a mode of life has had tremendous effects on education reform and reorganization for the interests of global capital (McLaren 2000; Bourdieu 1998). Neoliberalism as a political phenomenon involves the privatization of common resources for market/corporate interests, and is accomplished through a complex network of discourses, institutional arrangements, and ideologies that encourage self-investment. Neoliberal capitalism has grown beyond an economic-political system to encompass many aspects of human social life. Foucault's lectures on the *Birth of Biopolitics* describes this phenomenon and the rise of *homo economicus*: self-investing, entrepreneurial human beings responsible for their own success. An ideology of self-investment involves foisting all responsibility for survival onto students, rather than communities or governments (Simons 2006). The effects of neoliberalism in education are wide ranging, but overall they involve transitioning from valuing education as a public good, to the marketization of education for private interests—the effects of which are not equal for marginalized students of colour, students in poverty (see Lipman 2013; Rivera-Maulucci 2010). Science education has not been unaffected. Scholars have noted neoliberalism's negative effects on science for the public good (Bencze and Carter 2012) and school restructuring (Tobin 2011), as well as its role in steering education to meet the STEM human capital needs of particular corporations (Pierce 2013).

What should be the concern of educational research is how discourses of education work to (re)produce the subjectivities needed to maintain our current hegemonic, environmentally and socially destructive socioeconomic order (Hardt and Negri 2009). In this textbook analysis several themes emerged from the discourses around economic outlooks. For example, competition was described as a key element of science, over cooperation, and individuals were often positioned as the locus of action for the overall improvement of communities (Bazzul 2012). In addition, 'typical' discourses of careers compatible with an entrepreneurial focus where students were brought to invest in their own human capital while the sociopolitical backdrop of science work remains occluded, were found in these texts. I discuss the results related to neoliberalism and career discourses in relation to human capital more extensively elsewhere (Bazzul 2012, 2016).

The "Ethical Subject" of Science Education

The initial motivation for this research brief emerged when it became clear that these textbooks can also work to discursively constitute the possibilities for students and teachers to think and act along ethical lines in issues related to science. Examining how students and teachers are presented possibilities for ethical engagement requires specific theoretical frameworks from a Foucauldian perspective and involves thinking about "the way a human being turns himself into a subject" (Foucault 1982, p. 778), as well as specific relations to self, others, and the world required for ethical actions (Foucault 1988). Subjectification also consists of developing practices of self-examination, which influence a subject's actions toward others and the world (Peters 2004). Examining practices of self-formation is essential to a politics that reworks certain forms of subjectification (Butler 2004). I will discuss relations to self more closely in Chap. 5.

In Chap. 3 I will lay out in more detail how these Ontario biology textbooks delimit the possibilities for ethical thought and action in specific ways. In short, the discourses of these texts have students engage ethical issues on a juridical/policy level, consequently positioning many ethical issues under the umbrella of state governance. I argue that these texts work to constitute an ethical actor as someone who evaluates and amends policy and legislation and/or changes their personal life-style (Bazzul 2014b). In Chap. 4, I discuss how ethical issues operate along the poles of biopower in a society that is both disciplinary and regulatory. I position ethics in science (education) in terms of biopolitical engagement, or a 'push back' against biopower, along new ethical frontiers related to biotechnology, molecular genetics, etc. Thinking again about subjectivity, it is important to ask, 'who' discourses and practices of science education 'expect us to be' when we approach issues of ethical importance (Lather 2012). Chapter 3 is divided into two parts. Beginning with some of the results of this textbook study, I will cast a line of critical inquiry into ethics and science education discourses. I will then consider a broad political context that introduces the idea of relations of self that form the basis of ethical action.

Resistances and Reformulations

This textbook study helped contextualize the notion that discourses of education, in this case the discourses found in textbooks, shape our thinking and constitute subjectivities, whether we are aware of it or not. Subjectivities that are partly constituted by the discourses of science education are specific and can be traced, but also reworked and reformulated. Reworking these discourses, and the subjectivities they help constitute, is possible when educators endeavour to make the invisible visible through analytical engagement with discourses, and shift these discourses toward the promotion of an ecologically and socially just future.

Disruption of subjectification is always possible at the site where discourses and ideological practices that constrain and afford choices for students are repeated. Ironically, it is this site of repetition where resistance is possible because subjectivities, identities, and ways of being are sustained through constant, iterative performances at multiple sites. Taking the example of (the performance of) white(ness), male(ness), heterosexuality, able-bodied(ness) in educational settings, we must rework the repetitive practices and discourses that divide males and females, silence all other sexualities, and mask the Eurocentric, patriarchal image of science that becomes the assumed 'subject of science' (like the one alluded to in the epigraph to this chapter). Although power is exercised through discourses and their accompanying apparatuses (schools etc.), freedom is always possible because power can be "exercised over free subjects, and only insofar as they are free" (Foucault 1982, p. 790). Education is co-extensive with the production of subjectivities, and remains a site of resistance precisely because it can effect change. Subjectivity is central to modern struggles for justice, and in new informational, 'immaterial' economies, essential for social and political change.

Because people are attached to their subjectivities, reworking them is a radical project that can render people 'changed' (Butler 2004). As will be discussed in Chaps. 4 and 5, we may not have a choice. With growing social inequality and climate change unchecked, we may be reaching a point where we can no longer hold onto the subjectivities, beliefs, and worldviews that sufficed in the twentieth century.

As such, there is a need to reformulate how we relate to ourselves, the world, and others through science.

References

Althusser, L. (1998). Ideology and ideological state apparatuses. In J. Rivkin & M. Ryan (Eds.), *Literary theory, an anthology* (pp. 294–304). Malden: Blackwell.

Bazzul, J. (2012). Neoliberal ideology, global capitalism, and science education: Engaging the question of subjectivity. *Cultural Studies of Science Education, 7*(4), 1001–1020. doi:10.1007/s11422-012-9413-3.

Bazzul, J. (2013). *How discourses of biology textbooks work to constitute subjectivity: From the ethical to the colonial.* Doctoral dissertation, University of Toronto.

Bazzul, J. (2014a). Critical discourse analysis and science education texts: Employing Foucauldian notions of discourse and subjectivity. *Review of Education, Pedagogy, and Cultural Studies, 36*(5), 422–437.

Bazzul, J. (2014b). Tracing "ethical subjectivities" in science education: How biology textbooks can frame ethico-political choices for students. *Research in Science Education, 45*(1), 23–40.

Bazzul, J. (2014c). The sociopolitical importance of genetic, phenomenological approaches to science teaching and learning. *Cultural Studies of Science Education, 10*(2), 495–503.

Bazzul, J. (2016). Biopolitics and the 'subject' of labour in science education. *Cultural Studies of Science Education*, 1–14. (online first).

Bazzul, J. (in press). Biopolitics and the 'subject' of labor in science education. *Cultural Studies of Science Education.*

Bazzul, J., & Sykes, H. (2011). The secret identity of a biology textbook: Straight and naturally sexed. *Cultural Studies of Science Education, 6*(2), 265–286. doi:10.1007/s11422-010-9297-z.

Bencze, J. L., & Carter, L. (2012). Globalizing students acting for the common good. *Journal of Research in Science Teaching, 48*, 648–669.

Bhabha, H. K. (1994). *The location of culture.* New York: Routledge.

Blake, L. (2011). *McGraw-Hill ryerson biology 12.* Toronto: McGraw-Hill Ryerson.

Bourdieu, P. (1998). *On television.* New York: New Press.

Butler, J. (1990). *Gender trouble and the subversion of identity.* New York: Routledge.

Butler, J. (1995). Contingent foundations: Feminism and the question of "postmodernism". In S. Benhabib, J. Butler, D. Cornell, & N. Fraser (Eds.), *Feminist Contentions. A Philosophical Exchange* (pp. 27–35). New York: Routledge.

Butler, J. (1997). *The psychic life of power: Theories in subjection.* Stanford: Stanford University Press.

Butler, J. (2002). What is critique? An essay on Foucault's virtue. Accessed http://f-origin.hypotheses.org/wp-content/blogs.dir/744/files/2012/03/butler-2002.pdf

Butler, J. (2004). What is critique? An essay on Foucault's virtue. In S. Salih & J. Butler (Eds.), *The Judith Butler reader* (pp. 302–322). Malden: Blackwell.

Connelly, M., He, F., & Phillion, J. (2008). *The sage handbook of curriculum and instruction.* London: Sage Publications.

DiGiuseppe, M. (2003). *Nelson biology 12.* Canada: Nelson Thomson Learning.

DiGiuseppe, M. (2004). *Nelson biology 11: College preparation.* Toronto: Nelson Thomson Learning.

Dunlop, J. (2010). *McGraw-Hill ryerson biology 11.* Toronto: McGraw-Hill Ryerson.

Eco, U. (1989). *The open work.* London: Hutchinson.

Fairclough, N. (2003). *Analysing discourse: Textual analysis for social research.* New York: Routledge.

Fausto-Sterling, A. (2012). *Sex/gender: Biology in a social world.* New York: Routledge.

Foucault, M. (1970). *The order of things: An archaeology of the human sciences*. New York: Pantheon Books.

Foucault, M. (1972). *The archaeology of knowledge*. New York: Pantheon Books.

Foucault, M. (1977). *Discipline and punish: the birth of the prison*. New York: Pantheon Books.

Foucault, M. (1980a). *The history of sexuality. Vol. 1: An introduction*. New York: Vintage Book.

Foucault, M. (1980b). Two lectures. In M. Foucault & C. Gordon (Eds.), *Power = Knowledge: Selected interviews and other writings, 1972–1977* (pp. 78–108). New York: Pantheon Books.

Foucault, M. (1982). The subject and power. *Critical Inquiry, 8*(4), 777–795.

Foucault, M. (1984). What is an author? In P. Rabinow (Ed.), *The Foucault reader* (pp. 101–122). New York: Pantheon.

Foucault, M. (1986). *The care of the self (The history of sexuality, vol. 3)*. New York: Vintage Books.

Foucault, M. (2003a). So is it important to think? In P. Rabinow & N. Rose (Eds.), *The essential Foucault, selections from essential works of Foucault, 1954-1984* (pp. 170–174). New York: New Press.

Foucault, M. (2003b). What is critique? In P. Rabinow & N. Rose (Eds.), *The essential Foucault, selections from essential works of foucault, 1954-1984* (pp. 263–278). New York: New Press.

Foucault, M. (2003c). Questions of method. In M. Foucault, P. Rabinow, & N. S. Rose (Eds.), *The essential Foucault: Selections from essential works of Foucault, 1954–1984* (pp. 251–261). New York: New Press.

Hall, D. E. (2004). *Subjectivity*. Routledge.

Hardt, M., & Negri, A. (2009). *Commonwealth*. Cambridge: Belknap Press of Harvard University Press.

Irigaray, L., & Bové, C. M. (1987). Le Sujet de la Science Est-ll Sexué?/Is the Subject of Science Sexed? *Hypatia*, 65–87.

Jameson, F. (1991). *Postmodernism, or, the cultural logic of late capitalism*. Durham: Duke University Press.

Keller, E. F. (1996). The biological gaze. In G. Robertson, M. Mash, L. Tickner, J. Bird, B. Curtis, & T. Putnam (Eds.), *Future natural: Nature, science, culture* (pp. 107–121). London: Routledge.

Kendall, G., & Wickham, G. (1999). *Using Foucault's methods*. London: Sage Publications.

Kuhn, T. (1996). *The structure of scientific revolutions* (3rd ed.). Chicago: University of Chicago Press.

Lather, P. (2012). The ruins of neoliberalism and the construction of a new (scientific) subjectivity. *Cultural Studies of Science Education, 7*(4), 1021–1025. doi:10.1007/s11422-0129465-4.

Lipman, P. (2013). *The new political economy of urban education: Neoliberalism, race, and the right to the city*. Taylor & Francis.

Matthews, M. R. (2009). Science, worldviews and education: an introduction. *Science & Education, 18*(6–7), 641–666.

McLaren, P. (2000). *Che Guevara, Paulo Freire, and the pedagogy of the revolution*. Lanham: Rowman & Littlefield.

Mills, S. (1997). *Discourse: The new critical idiom*. London and New York: Routledge.

Ninnes, P. (2002). Discursive space(s) in science curriculum materials in Canada, Australia and Aotearoa/New Zealand. *Journal of Curriculum Studies, 34*, 557–570.

Peters, M. (2004). Educational research: 'Games of truth' and the ethics of subjectivity. *Journal of Educational Enquiry, 5*(2), 50–63.

Pierce, C. (2013). *Education in the age of biocapitalism: Optimizing educational life for a flat world*. New York: Palgrave Macmillan.

Rivera Maulucci, M. S. (2010). Resisting the marginalization of science in an urban school: Coactivating social, cultural, material, and strategic resources. *Journal of Research in Science Teaching, 47*(7), 840–860.

Roth, W. M. (2014). Enracinement or The earth, the originary ark, does not move: On the phenomenological (historical and ontogenetic) origin of common and scientific sense and the genetic method of teaching (for) understanding. *Cultural Studies of Science Education, 10*(2), 469–494.

Said, E. W. (1978). *Orientalism*. New York: Vintage Books.

Simons, M. (2006). Learning as investment: Notes on governmentality and biopolitics. *Educational Philosophy and Theory, 38*(4), 523–540.

Tobin, K. (2011). Global reproduction and transformation of science education. *Cultural Studies of Science Education, 6*, 127–142.

Chapter 3
The Ethical Subject of Science Education

Abstract In this chapter I present and discuss results from the textbook study that was introduced in Chap. 2. I demonstrate that discourses of Ontario biology textbooks position students to address particular ethical issues in very specific ways such as amending law and policy and making changes to personal lifestyles. I argue that discourses of science education inevitably limit the choices for students to engage ethically in science education and that educators need to explore other ways of thinking and acting ethically in science education, such as collective political action (protest and community organizing).

Keywords Ethics · Subjectivity · Discourse · Textbooks · Science education · Michel foucault

Ethics and Science Education

Science and science education are entangled with a wide range of ethical contexts ranging from the sharing of information and reporting, falsification/validation of data, use of animals, to the public return on investment (e.g. Is it ethical to allow people to sell organs? Is space-exploration a valid priority for the twenty-first century?). However, in terms of the wide-ranging problems facing populations, how does science education position communities to face the problems and issues that affect their everyday lived realities on a global scale? Who is science education meant to serve? What kinds of ethical engagement are, so far, difficult to achieve in typical school science? In this chapter I discuss one part of a textbook study with Ontario biology textbooks. That study specifically investigated how the discourses around ethical questions position students and delimit the choices they have when asked to ethically engage issues related to science. I argue that engaging ethics in science education should take a variety of useful forms. While educators can engage ethical debates delineated by those 'qualified' to have such debates (such as ethicists, government and scientific community representatives), they can also critically examine both the *topics* and *modes* of ethical engagement 'on offer' to students.

© The Author(s) 2016

J. Bazzul, *Ethics and Science Education: How Subjectivity Matters*,
SpringerBriefs in Education, DOI 10.1007/978-3-319-39132-8_3

Are urgent topics being covered? Who benefits from various forms of ethical engagement with issues related to science? In this chapter, and in Chap. 5, I argue for teachers to disrupt/change more typical ethical debates by suggesting different ethical questions and modes of engagement, even when they do not fit within what is already delineated by *sanctioned* government and academic discourses as relevant ethical engagement. Before moving on, it is important to recognize some of the work that has already been done in science education related to ethics.

Ethics in the Science Education Literature

In terms of addressing ethics directly as a topic in science education, scholars such as Zeidler (Zeidler and Sadler 2008), Reiss (1999), Levinson (2008), and Sadler (Sadler and Zeidler 2004) have laid significant groundwork by building frameworks for ethical engagement. In addition to this work, many science educators have placed social and environmental justice and activism at the heart of their work, giving it a strong ethical dimension (Weinstein 2008; Mensah 2009; Tolbert 2015; Dimick 2012; Bencze and Alsop 2014; Emdin 2010). In his book on scientific literacy, Hodson (2008) dismisses the claim that simply by learning about science as the courageous pursuit of 'truth' outside of personal interests, for example by learning/ following Merton's (1973) norms, students will naturally become more ethical. Instead, Hodson claims that "science would benefit from the transfer of ethical standards in the opposite direction" (p. 12). Reiss (1999) provides guidelines for teaching ethics in science education. These include criteria for coming to valid ethical conclusions, such as the internal consistency of argumentation, the relation between arguments and already existing ethical frameworks, and whether an ethical view-point achieves validity and consensus through debate. While this approach to ethics in science education is very useful, it is also quite specific in its focus on logic, validity, and consensus. Some important aspects of ethical engagement may be missing from an approach reliant on logic and debate. Critical educators are aware that ethical actions consist of many practices and contexts that arise (and arrive!) in classrooms through emotional trauma, spirituality, migration experiences, and community needs. Though there are always assumed approaches to ethical issues related to science, these approaches can be remolded by engaging other knowledges already held by students, such as indigenous ways of understanding the natural and social world.

Here I am consciously trying to re-approach ethics and science education, much like Blades (2006) does in his 'encounters with the eroticism of nature' through the philosophy of Emmanuel Levinas. Blades' (2006) essay, "Levinas and an ethics for science education", makes the critical point that little to no attention has been given to theorizing ethical frameworks for science curriculum and pedagogy. Regarding the mission of STS(E) education (a Canadian usage meaning Science, Technology, Society, and Environment education), Blades maintains that,

... the nature of ethics and responsibility essential to this mission are not considered in the theorizing of an STS(E) approach; there is no discussion on how responsibility informs action, such as how to understand whether students should engage in dissection, or how to approach the natural world, such as [the dissection of] frogs, from an ethical perspective. (p. 648).

Blades proposes an extension of Levinas' (1979) concept of 'other' as a basis for ethics in science education by extending it to include non-human and abiotic entities. Whether or not one agrees with Levinas' ethics, Blades addresses the question of ethics in science education on a philosophical level. This move is radical in that it works "to change the very concepts of the debate" (Žižek 2009, p. 51). In reformulating ethics as a radical, erotic, and terrifying relationship to the other, Blades gives science educators an alternative for ethical engagement.

A particularly good overview of ethics, and its relation to socioscientific issues (SSI), is Saunders and Rennie's (2013) paper "A Pedagogical Model for Ethical Inquiry into Socioscientific Issues In Science". This paper provides an overview of some of the ways ethical engagement has been conceived thus far in relation to the lived realities of students and other social issues. What is particularly useful about their literature review is that they pull from a diversity of scholars as well as curricular and pedagogical resources. Indeed, a good place to begin speaking about ethics is the incorporation of controversial issues in science education. The *Socioscientific Issues* (SSI) research paradigm [see Sadler's (2011) book on SSI specifically], looks at the integration of controversial real world issues. This movement is based on the notion that science education has a responsibility to society, and that students must understand this responsibility and their ethical place (Van Rooy 1993; Reiss 2008).

Lock and Ratcliffe (1998), who arguably helped re-introduce ethics into science teaching in the United Kingdom, tie a *separate* conception of ethics to the social in the *ASE Guide to Secondary Science Education* (UK) under the heading "What do We Mean by Social and Ethical Applications of Science":

Social applications of science are those that impinge on our lives directly everyday and indirectly through political and economic decision-making. Science provides us with the evidence for what we can do, whether it be cloning, 'splitting' the atom or making new chemicals. Science is a process of rational enquiry that seeks to propose explanations for observations of natural phenomena. Ethics helps us to decide what we should do. Ethics is a process of rational enquiry by which we decide on issues of right (good) and wrong (bad) as applied to people and their actions (p. 110).

This passage makes three important points. First, it gives an *explicit* role for ethics in science education—science is a process of rational enquiry into phenomena, and ethics is the rational enquiry into right and wrong—'what we should do'. However, scholarship in fields such as science and technology studies, the history and philosophy of science, and critical feminist studies, has shown that science and ethics are hopelessly entangled. That is, ethics is already an *inseparable* part of: (i) the attitudes and core practices of scientists related to designing and carrying out research; (ii) how scientists share information and work with each other and community partners; (iii) how decisions on research and funding are made at the level of governance and community participation. If we now include the cultural and sociopolitical realities

of education, what is worthwhile to learn, whose knowledge and stories are important, how the social world is described and materially distributed, we find that the
'subjects' of ethics and science cannot be divorced. At every turn, science involves
ethics—what we should do. Furthermore, schooling/science education produces
constituted subjects who understand their subjectivity in terms of their social,
cultural, and political milieu. That is, subjects who perform the identities of science
student/scientist/citizen/activist/change-maker; subjects who are simultaneously
'ethical subjects'.

Even more importantly, Lock and Ratcliffe (1998) make a crucial step in bringing
ethical concerns into science. They correctly and succinctly define ethics in its most
basic form, "[deciding] on issues of right (good) and wrong (bad)" (p. 110), or more
basically *what we should do*. Viewing ethics in its most basic form gives analysts a
more wide-ranging, comprehensive scope of what comprises ethical issues, decisions, actions, and the subject of these decisions and actions. In this way critical
scholars can analyse the discourses of science education for the many dimensions
of ethics already present and at a play. This is to say that, it is not enough to bring
ethics to science education. Educators must engage in critical-analytical tasks to
realize what is *actually* presented to students in terms of ethical practices and possibilities. In the following sections I draw from the textbook study outlined in the last
chapter to give an example of such a critical exercise. Being critical of the opportunities given to science teachers and students to 'be ethical' makes it easier to put
forward ethical frames that politically engage social and environmental issues.

Ethics, Textbooks, and Curriculum

Now that we have laid some groundwork for thinking about subjectivity and
approaching textbooks using a Foucauldian discourse analysis methodology, I want
to consider how discourses of science education can work to constitute an ethical
subjectivity. I will do this by discussing some results from the textbook study
outlined in Chap. 2 in terms of how students are brought to ethical exercises and
questions in Ontario biology textbooks. I revisit some of the methodological aspects
discussed in Chap. 2 to better situate this examination of ethics in education. Also,
the chapters of this research brief are independently searchable in databases and
therefore some context is needed for new readers. These results have been published
in the journal *Research in Science Education*, which includes details not outlined in
this chapter (Bazzul 2014). I argue that discourses found in the Ontario biology
textbooks in question constrain how students are brought to ethical questions and
exercises, as well as possible actions students may take. These constraints help to
produce a kind of 'ethical subjectivity'; how students take up the role of, and see
themselves as, ethical actors and decision makers.

One of the key points I want to make is that it is not just any ethical question,
action, or position that can be taken up within the discursive frame of this text. As I
demonstrate below, choices for ethical decision making revolve around themes of

policy recommendation, health, and personal lifestyle choices. While these themes are relevant, they represent a limited set of ethical concerns and modes of action. To address this specificity, it is useful to contextualize how ethics is situated in science education from the perspective of governance and power. To this end, I will introduce Foucault's concepts of biopower and biopolitics as a broad way to think about how it has come to be that certain modes of ethical engagement are prominent in state education. In short, this involves how populations come to be governed in modernity, with schooling being a key site of this governance. A shift in perspective is necessary because current modes of ethical engagement are seemingly inadequate for dealing with growing social inequality and ever-greater environmental destruction. Science education plays a big role in shaping how educational communities can ethically engage.

Discourses of science education, in this case biology education, work to constitute the kinds of subjectivities 'required' for science work, research, and everyday decision-making. Critical examination into discourses surrounding ethics can help rework how ethical issues are framed in curriculum, what constitutes ethical action, and 'who' curriculum expects ethical actors to be. Science and ethics may be too focused on juridical questions, such as those found in bioethics. I agree with Foucault (2003) in that "contemporary political thought allows very little room for the question of the ethical subject" (p. 294). This chapter's main goal is to show the specificity of ethical engagement in science education. In Chap. 5, "Political Engagement as a backdrop for Dimensions of An Ethical Self" I will explore further the relations of self (self to other, self to self, and self to the world) of this ethical subject. These dimensions of self may be necessary for twenty-first century ethical engagement that works towards the goals of social and environmental justice.

The remainder of this chapter consists of two basic sections. The first discusses some examples of directly ethical questions and exercises from the four Ontario biology textbooks. Again these are: *Nelson Biology 11 College Preparation* (DiGiuseppe 2004); *Nelson Biology 12* (DiGiuseppe 2003); *McGraw-Hill Ryerson Biology 11* (Dunlop 2010); *McGraw-Hill Ryerson Biology 12* (Blake 2011). The second section situates the recurring themes of the ethical questions and exercises of these textbooks: policy reform, population health, individual responsibility, and government regulation, in relation to Foucault's historical work on governance, biopower, and biopolitics (Foucault and Senellart 2010; Foucault et al. 2007).

Continuing with a Foucauldian Methodology

The methodology for looking at these texts was outlined in the previous chapter; however I'd like to detail, and reiterate, a few key points. First, discourse consists of statements that, having varying amounts of institutional support, have effects on how people think and act (Foucault 1972; Mills 1997). This study related this conception of discourse to ethical exercises and questions in these textbooks in order to examine how choices for ethical engagement are delimited, and what kinds of ethical subjectivities these textbooks may help constitute. Questions and exercises that have students decide on a "right" course of action comprised the discursive data for this analysis, as they work to position students as directly ethical actors.

The idea of isolating instances in the text where students decide on a right or wrong course of action came, in part, from Weber's (Weber and Andreski 1983) distinction between the *tenets* and *practices* of religions—where the two can often operate with different rationales, motives, and goals. Focusing explicitly or directly on ethical exercises and questions separates the tacit ethical messages/contexts (e.g., the description of a social issue) from the instances where students actually, directly decide on a course of action. It is true that an entire text does much to influence how students engage ethically, however looking at specific instances where students are meant to think and act according to a 'right course of action' can give educators a more detailed picture of the choices afforded to students around ethical decision-making and ethics in general. A second reason to focus on these specific instances is that an archeological analysis tries to concentrate on the surface or literal meaning of statements, without inferring a meaning "under the surface." While textbook authors have intentions, what is most salient in an archeological analysis are the actual ways statements operate and convey relationships found within the discourses of these texts, the limits statements set, and the affordances they provide—all of which contribute to the constitution of particular kinds of subjectivities (in this case ethical subjectivities). As stated in Chap. 2, the four Ontario biology textbooks used in this study represent 'sanctioned knowledge', under the labels of both science and government. This government-sanctioned aspect is relevant when we begin to consider how students come to engage ethically in some ways but not in others. That is, in ways that align with the interests of governments. While these textbooks were originally examined in terms of how they could work to constitute various kinds of sex/gendered, colonized/colonizer, and political–economic subjectivities (see Bazzul 2012; Bazzul and Sykes 2011), two basic patterns emerged related to ethical engagement.

First, teachers and students were seldom led to think or act on a *directly* ethical level. The term 'directly ethical' should be understood broadly as any instance where a student is meant to deliberate about a right behavior, right course of action, or what individuals and groups should do, based on either an implied sense of responsibility or explicit obligation. Again, though a wide range of discourses inform ethical behaviours and actions, trying to isolate those situations that have students and teachers come to make ethical decisions, or take actions, allows educators to examine the discursive limits of ethical decision-making, thought, and action.

The second pattern to emerge was that the choices made available to students to engage ethically were constrained in particular ways. This led me to ask, 'what constitutes a legitimate ethical issue?' and 'what are the range of choices offered to students when taking ethical action?' Once exercises and questions where students were directly asked to think and act along ethical lines were isolated, the discursive data was analyzed according to the types of actions students were given as choices, whether or not they were meant to actually carry out these actions. Categorizing all ethical questions and exercises was impossible; however, some recurring themes emerged. In general, ethical engagement involved the following: regulation of a wide variety of practices and substances from research funding to drugs; altering personal lifestyle choices; policy and legislative/legal reform; and optimizing the health of individuals and/through the health of the population.

What Qualifies as an "Ethical Question or Exercise" in These Texts?

What constitutes a 'directly ethical' question or exercise is necessarily problematic. In this analysis, and also in general terms, I advocate for a very basic definition of ethics as Lock and Ratcliffe (1998) do. The Encyclopedia of Postmodernism (Taylor and Winquist 2002) defines ethics as "the historical inquiry into how one is to be" (p. 114). According to Merriam-Webster's online dictionary, "ethics" is defined first of all as "the discipline dealing with what is good and bad and with moral duty and obligation" (Ethics n.d.). Therefore, I chose to identify directly ethical exercises/

Table 3.1 Exercises and questions that fit, or did not fit, criteria of being 'directly ethical' (from Bazzul 2014)

Criteria for directly ethical questions	Examples that qualify	Examples that do not qualify (with explanation)
Students are directly asked to take a position on an issue	"(Debate statement) The government should allow xenotransplants in Canada" (DiGiuseppe 2003, p. 360)	"Are organisms more than just their genes?" (Blake 2011, p. 372) (*No direct backdrop of obligation or responsibility; "correct" course of action not a concern*)
This position comprises something that "should" be done; a "correct" course of action	"Write a brief position paper containing your recommendations to a legislature on (the decriminalization or legalization of marijuana)" (DiGiuseppe 2004 p. 183)	"Taking a global view (societal, economic, etc.), list the advantages and disadvantages of UHT technology" (DiGiuseppe 2004, p. 115). (*Exercise does not directly engage a "correct" course of action; dimension of responsibility is not explicit*)
The position, decision, or action taken is related to an implicit or explicit responsibility or obligation of individuals and/or organizations	"(In terms of treatment or prevention of a disease) What do you think the priorities for research should be" (Dunlop 2010, p. 460)?	"What steps could people who are allergic to fungal spores take to ensure that their living environment is relatively spore free?" (DiGiuseppe 2004, p. 140)? (*Question lacks both a dimension of responsibility and a chance to deliberate on a correct course of action*)

Table 3.2 Recurring ethical themes and examples (Bazzul 2014)

Recurring theme	Examples from four Ontario biology textbooks
Policy recommendation at government level	"Debate with classmates whether Canada should adopt an 'opting out' policy to increase the number of cadaveric donors. What problems might this create" (Blake 2011, p. 119)?
Research funding decisions, government or otherwise	(In terms of treatment or prevention of a disease) "What do you think the priorities for research should be" (Dunlop 2010, p. 460)?
Regulating the use of particular substances	"Do you think people's urine should ever be tested for drugs? Use a PMI chart to examine the advantages and disadvantages of urine testing and make recommendations on its use" (DiGiuseppe 2004, p. 214)
Considerations of specifically regulating genetic research/biotechnology	"What effects might the creation of such 'designer babies' have on society? Explain what laws, if any, you think the government should enact to regulate this area of genetic research" (Blake 2011, p. 329)
Optimizing health of population	(Debate statement): "Women should not drink even small amounts of alcoholic beverages while pregnant" (DiGiuseppe 2003, p. 120)
Personal habits/lifestyle choices	"One of your friends talks constantly about losing weight, while another has decided she is going to become a vegan. What advice would you give each of them, and why" (Dunlop 2010, p. 366)?
Regulating food safety at government level	"Nutriceuticals should be regulated as drugs under the Canada Food and Drug Act" (DiGiuseppe 2003, p. 5)
Using cost/benefit analysis (pros/cons)	"Your town council is debating whether or not to pass a resolution banning pesticide use in local parks. Compile a list of pros and cons that could be used in reaching a decision" (DiGiuseppe 2003, p. 144)

questions as ones where students are asked explicitly to recommend or defend a "right behavior" or "correct course of action" in relation to explicit or implicit responsibilities of individuals and organizations. Such identification helps target those discursive instances where students are put in the position of ethical actor. While everyday pragmatic decisions and unusual ethical dilemmas can be imagined under this rubric, this is exactly the point! These decisions and dilemmas would have *also* been included as part of ethical discourse. This definition of ethics can be employed in an analysis to gather any theme surrounding directly ethical exercises and questions. It should not be confused with a comprehensive approach to ethics where ethics is defined specifically, for example as a commitment to social or ecological justice, or what it may mean to produce reliable scientific research.

Table 3.1 lists the criteria by which directly ethical questions and exercises were selected, including examples of questions/exercises that did and did not qualify.

So while questions/exercises concerning endangered species or summits against environmental degradation obviously have ethical dimensions, this analysis only included such exercises if they required students to directly think and act concerning a 'correct', 'right', or 'responsible' course of action. An ethical question/exercise and a possible response can be seen as separate, yet also inclusive of each other, in that the context of an ethical question often sets the limits for its solution. In a sense, looking at directly ethical questions and exercises is a discursive limitation that, in turn, exposes limitations concerning choices for students to make ethical decisions and/or act in an ethical manner. Isolating directly ethical questions/exercises in these textbooks yielded about 15–25 per textbook; Table 3.2 presents some examples according to eight themes that emerged around ethical exercises/questions.

Analysis of Ethical Questions and Exercises in Biology Texts

After examining directly ethical questions and exercises, six recurring themes emerged in all four textbooks and two recurring themes in two of four and three of four textbooks respectively. These themes are related to the kinds of actions students could take in response to these ethical questions and exercises within the discursive frame of the texts. Table 3.3 summarizes these themes and the textbooks in which they appear. This is not to say that these themes are the only ones surrounding directly ethical questions and exercises; rather they represent just one possibility for characterizing the choices offered to students.

Table 3.3 Recurring themes of directly ethical questions/exercises in Ontario biology textbooks (Bazzul 2014)

Recurring theme	Texts where themes are found
1. Policy recommendation at government level	✓ McGraw Hill biology 11
2. Research funding decisions, government or otherwise	✓ McGraiw Hill biology 12
3. Regulating the use of particular substances	✓ Nelson biology 11: college preparation
4. Considerations of specifically regulating genetic research/biotechnology	✓ Nelson biology 12 (All four textbooks)
5. Optimizing health of population	
6. Personal habits/lifestyle choices	
7. Regulating food safety at government level	✓ Nelson biology 12
	✓ McGraw Hill biology 11
	✓ McGraw Hill biology 12
8. Using cost/benefit analysis (pros/cons)	✓ Nelson biology 12
	✓ McGraw Hill biology 12

From these tables we can see that whenever students are led to directly think and/ or act ethically, they very often must engage in terms of juridical/legal concerns, such as regulating certain practices and substances, making policy recommendations, deciding on what scientific research should be funded, ensuring both personal and population health, and altering personal habits/lifestyle choices. These themes are interconnected, and I'd like to contextualize them in the following sections in terms of Michel Foucault's notions of governance, biopower, and biopolitics.

Contextualizing Ethical Themes in Textbooks: Governance, Populations, and Lifestyle

The themes surrounding ethical exercises and questions can be contextualized using Foucault's historical analyses concerning biopolitics and the governance of populations. In this section I will address the recurring 'ethical' themes of: government and policy; health and population(s); and the individual and lifestyle choices in the context of biopolitics, subjectivity and bodies. The discussion around biopolitics will be continued in Chap. 4.

One aspect that unites these interrelated, recurring themes of health, population, regulation, and policy concerns regarding ethical questions and exercises in these textbooks, is that they fall, more or less, within the purview of state governance. An ethical actor, according to these texts, is one who largely endeavours to change, evaluate or amend law and/or policy. It seems commonsensical that government-funded science education would have students engage ethical scenarios that are also the concern of government (e.g. regulation, personal care, policy change, safety, health). However, critical educators should also ask how these concerns, and why these modes of engagement, come to be a focus of education (and not others). Taking a biopolitical approach, which consists of taking the 'modes of life' (at the level of populations) as the object of politics, can help educators make sense of how certain modes of engagement are favoured over others in state education. So, to summarize the main points concerning ethics thus far:

- Not any ethical issue or course of action is afforded to students because science education discourses limit the range of possibilities for ethical thought and action.
- The ethical contexts and possibilities afforded by science education discourses appear to fall under the purview/viewpoint of government concerns.
- Educators, for the sake of ecological and social wellbeing, must think beyond the range of ethical actions and decision-making afforded by science education discourses
- Pushing against science education discourses is biopolitical, that is, a process of political struggle against the exercising of (bio)power through institutional networks.

The recurring ethical themes and modes of engagement outlined in these texts are not wrong or dangerous. Rather, the goal is to improve ethical engagement in science and biology education through an understanding of its limitations and motivations—ones that exceed the will of any individual or group of textbook writers—toward understanding and shaping what kind of ethical subjectivities are important in science curriculum. What kind of people, citizens, scientists, community members does curriculum value and nurture? Before we move ahead, let us take a closer look at some of the recurring themes in this analysis.

Government Regulation and Policy Recommendation

Each of the four textbooks discursively positioned students to deliberate on government policy reform, regulation, and amendments to law. This places ethical decision-making and action within a legal–juridical context because students must consider what laws and policies are already in place, making a legal/juridical solution a large part of the framing of an ethical issue. Here, an ethical subject must make an ethical decision in total accordance with the law; either what laws are already in place, or those yet to come.[1] The law therefore gives legitimacy to ethical modes of action prior to a student's engagement. Consider the following two examples:

1. "Should labeling be required by federal and provincial governments?" (Dunlop 2010, p. 574)
2. (Debate statement): "The government should allow xenotransplants in Canada" (DiGiuseppe 2003, p. 360).

In both examples students must consider the 'right' thing to do from a government policy perspective. The textbook rationale for labeling in the first example is so that consumers can make good decisions about genetically modified organisms (GMOs) in food (Dunlop 2010, p. 574). Here the ethical question is restricted to debate about extending an already existing labeling policy to GMO foods. That is to say, it is not a debate about the availability of GMO or GMO-free foods, or whether GMO foods are ethical, or the ethics of starvation in the Global South, etc. Though the labeling debate could address these topics, students grapple with the technicalities of labeling laws—not whether, or how, current agricultural practices punish and subjugate the 'less developed' world. The implication is that students and teachers may come to see ethical action as centered around legalities for consumers. More important topics, such as a reliance on market forces to guide social policy, may function as the ideological backdrop of such exercises, yet they are not open for debate. In this case, a rather close-ended approach underlines the painful assumption that multinational corporations are not expected to change packaging procedures on their own. In the

---

[1]This is a literal example of Butler's (1997) point that a subject's freedom is tied to subjection to the law. Butler (1997) does not mean only "law" in the juridical sense, but any social order that constitutes individuals as subjects.

xenotransplant example, the student is led, not to examine cultural beliefs about the meaning of non-human life, but to reason from the standpoint of government policy about a health-related issue. Ethical action and decision-making become tied to the economics of livestock and population health in relation to disease.

These ethical exercises give legitimacy to the state's concerns, but simultaneously make the state 'disappear' as there is little context for making ethical decisions, or taking ethical action, outside a legal–juridical context. What if teachers and students wanted to speak about ethical problems in which governments were culpable? These problems could include the destruction of indigenous environments and knowledges, collusion with private interests over the public good, or the pursuit of economic dominance at any cost. A strong interconnectedness between government policy and science makes sense because governments make decisions about research according to the legitimacy, infrastructure, technology, and labour that science provides government. Approved science textbooks in Ontario must cover at least 85 % of the Ontario curriculum (Ontario Ministry of Education 2008), which, in a sense, also positions them as government documents. It is easy to imagine a situation where 'objective' statements being spoken under the authority of government or science could become interchangeable in educational settings, especially when textbook authors are anonymous.

Health and Populations

Ethical questions regarding personal health are linked to questions of regulation and law/policy reform as well as concerns about population health. The following three examples fall under the recurring theme of optimizing the health of populations.

1. (Debate statement): "Women should not drink even small amounts of alcoholic beverages while pregnant" (DiGiuseppe 2003, p. 120).
2. "Given the potential benefits and risks, prepare a brief report arguing whether or not the use of artificial blood should eventually replace natural blood in all cases" (Dunlop 2010, p. 298).
3. (Debate statement): "Non-therapeutic use of antibiotics in farm animals should be banned" (DiGiuseppe 2004, p. 127).

Foucault argues that questions of health, regulation, and policy, from food safety to farming practices, are integral to the governance of individuals and populations in modern western societies (Foucault et al. 2007). His historical analyses outline a transition in governance from corporeal forms of discipline, to the employment of law, 'policy', and statistics that emerged when 'the population' became a political problem that had to be managed by governments (Foucault et al. 2007). As 'the population' began to be seen as a 'natural' phenomenon in the seventeenth and eighteenth centuries, this meant it also became an object of health management. As Foucault (Foucault et al. 2007) explains:

> ... the management of this population required, among other things, a health policy capable of diminishing infant mortality, preventing epidemics, and bringing down the rates of endemic diseases, of intervening in living conditions in order to alter them and impose standards on them (whether this involved nutrition, housing, or urban planning), and of ensuring medical facilities and services (p. 367).

Populations, their modes of social life, productivity, pathologies, and purposes become the focus of various technologies of control, or *biopower*, including scientific knowledges, that combine with these technologies of control to help govern, regulate, divide, subjectify, as well as inform.

The population becomes something that can be 'intervened upon', from economic activities to the attitudes of subjects.

> The final objective is the population. The population is pertinent as the objective and individuals, the series of individuals, are no longer pertinent as the objective, but simply as the instrument, relay or condition for obtaining something at the level of the population. (p. 42).

This population becomes "the reality the state will have to be responsible for, rather than individuals who must be subjugated and subject to imposed rules and regulations" (p. 352). This involves managing hygiene and resource scarcity, as well as developing demography and economy measurements for the 'natural' phenomenon of the population. One of Foucault's overall points is that the notion of the 'population' is also an historical contingency, even though it is a tangible noun in the parlance of our times. It emerges alongside institutions of control, including educational institutions and 'scientific' discourses that work to shape the conduct of subjects.

Understanding biopower, the control of populations and the harnessing of their productive power, relies heavily on knowledges that recast human beings as a species:

> By this, I mean a number of phenomena that seem to me to be quite significant, namely the set of mechanisms through which the basic biological features of the human species become the object of a political strategy or a general strategy of power or, in other words, how, starting from the eighteenth century, modern Western societies took on board the biological fact that human beings are a species (Foucault et al. 2007, p. 1).

This is not to say that human beings are *not* a species, but only that this knowledge is linked to practices of governance. Foucault links the population to its roots in biology; entwining science and politics: "The population is therefore everything that extends from biological rootedness through the species up to the surface that gives one a hold provided by the public" (p. 75). Foucault establishes a relationship between objective knowledge, from biology to economics, and governance:

> You have a science which is, as it were tête-à-tête with the art of government, a science that is external to the art of government and that one may perfectly well found, establish, develop, and prove throughout, even though one is not governing or taking part in this art of government. But the government cannot do without the consequences, the results, of this science. So, as you can see, a quite particular relationship of power and knowledge, of government and science appears (Foucault et al. 2007, p. 351).

Science textbooks that are written under the supervision of modern governments therefore can be understood, to some extent, as interventions for the management of populations. In this way, the recurring themes around ethical questions and exercises in these biology textbooks can also be seen as historically contingent. Therefore, educators can view ethical orientations and themes in science education materials as malleable and always open to change.

Constituting an ethical subject through limitations and affordances in discourse requires repetition. Butler (1997) is clear that, "if conditions of power are to persist, they must be reiterated; the subject is precisely the site of such reiteration, a repetition that is never merely mechanical" (p. 16). Educators can examine both the discursive limitations and affordances in curriculum materials and also intervene at the site of where students have chances to engage in ethical action and/or decision-making. What ethical actions are open to students and how can educators rework and expand the foci for ethical engagement?

The Individual and Lifestyle Choices

Ethics can also be associated with personal lifestyle choices. Foucault's (Foucault 1985, 1986) work shows that ethical concerns are often refracted through relations of self, which I will discuss further in Chap. 5. The following three examples highlight how directly ethical questions in these textbooks can be framed within a context of individual choices and lifestyle.

1. (What students can do to defend the biosphere) "Take action. Decide what issue is most important to you and pursue it. Write to politicians and corporations expressing your opinion" (Dunlop 2010, p. 608).
2. (Preamble to question about the willingness to change personal behavior) "Individuals can do little about the decisions of governments and corporations; they do, however, have control of their personal lifestyle decisions" (DiGiuseppe 2004, p. 423).
3. "List three changes that can be made to your personal lifestyle that would reduce the odds of a mutation taking place" (DiGiuseppe 2003, p. 263).

According to example 2, since individuals 'cannot do much against governments and corporations', they are encouraged to take action individually. In these examples, an ethical subject is meant to see the locus of ethical action in personal and lifestyle choice. Peaceful dissent, community organizing, signing petitions, and civil protest, do not seem to be choices for students. The absence of options for collective action—remembering that discourse also involves what is not said—is another way students are discursively oriented towards individual ethical actions and behaviour.

Foucault's historical analyses reveal that after populations become the focus of political power, comes the co-development of a self-governing individual (Foucault and Senellart 2010). Foucault's notion of *governmentality* refers both to structures and institutions that overtly govern and, more broadly, to any discourse or practice that 'conducts the conduct' of others and one's self (Lemke 2011). The legitimacy of modern nation states after 1945 rests in their ability to allow a 'competitive market' to function while certain 'freedoms' become less important (Foucault and Senellart 2010). Increasingly, in neoliberal political systems, we find the locus of ethics and politics residing at the level of the individual. Under neoliberalism, the individual is not just a rational unit of liberal economics, whose self-interest is essential to the functioning of capitalist markets, but one who must become a competitive entrepreneur–with state education promoting constitution of this neoliberal subjectivity (Simons 2006). Are students constituted as ethical actors who see 'themselves' as the sole locus of action and transformation? In the absence of collective action, these ethical questions could be seen to be co-extensive with (neoliberal) governmentality through state-sanctioned biology textbooks. As Lemke (2011) points out, to understand governmentality, the individual's role in maintaining a certain, entrepreneurial, self-centered political order must become the target of political, social and cultural analysis.

Thinking About Relations to Self

A student's conception of self, and the relationship of this self to others and the world, also plays a role in the constitution of an ethical subjectivity. In his analyses of the classical world, Foucault (1986) describes how these relationships, exercised through practices of self, are important to consider alongside other ethical codes:

> … It is easy to conceive of moralities in which the strong and dynamic element is to be sought in the forms of subjectivization and the practices of self. In this case, the system of codes may be rather rudimentary. Their relative observance may be relatively unimportant, at least compared with what is required of the individual in the relationship he has with himself, in his different actions, thoughts and feelings as he endeavors to form himself as an ethical subject. (p. 30)

In becoming an ethical subject, rules and codes are still important; however, it is important to consider "how the individual establishes his relation to the rule and recognizes himself as obliged to put it into practice" (p. 27). Thinking about the specific relationships of self to other, self to self, and self to the world can help us better understand what may appear to be similar ethical actions and positions, but embody completely different modes of 'being an ethical subject'. Self-formation as an ethical subject involves:

> … a process by which the subject delimits that part of himself that will form the object of his moral practice, defines his position relative to the precept he will follow, and decides on a certain mode of being that will serve as his moral goal…Moral action is indissociable from these forms of self-activity. (Foucault 1985, p. 28)

Practices of self also need to be examined when looking at how ethical subjects are meant to act in science education. I shall discuss these relations in Chap. 5, with a view to making democratic, egalitarian politics integral to these relations in an effort to meet the major challenges of climate change and social inequality.

Toward Different Approaches to Ethics in Science Education

While the regulation of various practices, legal and policy reform, and lifestyle choices, are not negative options, it is questionable whether they alone are appropriate when contending with what Žižek (2009) identifies as the four great antagonisms of the twenty-first century: (1) The inappropriateness of intellectual property that should be made common; (2) The ethical implications of biotechnology and genomic research; (3) Climate change and ecological collapse (4) New forms of apartheid, and 'human zoning'. What role can science education have in facing these challenges? Ethics in biology and science education must deal with new problems, such as the way biotechnology has recast discourses regarding human life at the molecular level (Gerlach 2011). Educators can ask after the subjective positions concomitant to modes of ethical engagement and knowledge. I will conclude this chapter by briefly discussing specific topics that require ethical engagement in the coming decades.

Biotechnology, Medical Research, and New Considerations for Life

Rose's (2007) work establishes important relationships between political, social, and economic contexts and biomedical research. His work raises the following questions that should inform a conversation around ethics and science/biology education:

1. **Private interests**: How do private interests shape research? How does the narrative of personal economic 'success' shape the way we see ourselves living 'productive' lives?
2. **Genomic research**: How does genomic research shape how we understand ourselves in relation to disease and genetic enhancement? How is 'self-responsible behavior' tied to risk and self-enhancement? How is ethics linked to opportunity and hope?
3. **Governance**: How does biological knowledge order priorities for people and organizations? How do social movements influence scientific knowledge production?
4. **Nonhumans**: 'Who' has a 'right to life'? How can educational communities breakdown biotic/abiotic, human/animal, and included/excluded binaries?
5. **Gatekeepers and principles**: Does biotechnology require new ethical gatekeepers (e.g. counselors, researchers, and policy makers)?

Politicizing Questions of Environmental Destruction

Although climate change and topics such as deforestation find their way into curriculum, the choices offered to students and teachers for ethical engagement need to effect change. The following suggestions may help provide options for engaging environmental issues of importance.

1. **Organized action**: Provide options for organizing communities such as time and web-based resources.
2. **Multi-faceted solutions**: Issues of social and ecological justice require many comprehensive responses. Many immanent problems have solutions that have yet to come.
3. **Freedom to challenge**: Provide the freedom to challenge assumptions underlying ethical decision making (e.g. distinctions between "humans" and "nature"). Who gets to make decisions about key issues, and who does not? Are there other ways of engaging ethical issues?

While these are general suggestions, they may help challenge assumptions and afford more choices for students to engage ethically through science education. This study has identified themes surrounding directly ethical questions and exercises in biology textbooks related to health, individual and lifestyle choices, and policy/legal reform that may work to discursively constitute a very particular kind of ethical subjectivity in students. In Chap. 4, I lay out a political context that can inform different ethico-political approaches to twenty-first century issues.

The goal of this chapter was to demonstrate that discourses of science education already work to constitute 'ethical subjectivities' through specific affordances and limitations in relation to the types of ethical contexts and modes of engagement offered to students. I also suggest that these modes of engagement are not enough to meet the challenges of the twenty-first century. What is also needed are more investigations into the way we come to see ourselves as ethical actors through education. I believe this 'ethical subjectivity' needs to be understood in a biopolitical context, and guided by egalitarian, democratic politics. At stake is how students and teachers come to see their lives through institutions and practices of education. In Chap. 5 I will argue that a more overt emancipatory politics should inform the dimensions of ethical subjectivity through relations to self, other, and the word.

References

Bazzul, J., & Sykes, H. (2011). The secret identity of a biology textbook: Straight and naturally sexed. *Cultural Studies of Science Education, 6*(2), 265–286. doi:10.1007/s11422-010-9297-z.
Bazzul, J. (2012). Neoliberal ideology, global capitalism, and science education: Engaging the question of subjectivity. *Cultural Studies of Science Education, 7*(4), 1001–1020. doi:10.1007/s11422-012-9413-3.

Bazzul, J. (2014). Tracing "ethical subjectivities" in science education: How biology textbooks can frame ethico-political choices for students. *Research in Science Education, 45*(1), 23–40.

Bencze, L., & Alsop, S. (Eds.). (2014). *Activist science and technology education*. Dorecht, NL: Springer.

Blades, D. W. (2006). Levinas and an ethics for science education. *Educational Philosophy and Theory, 38*(5), 647–664.

Blake, L. (2011). *McGraw-Hill ryerson biology 12*. Toronto: McGraw-Hill Ryerson.

Butler, J. (1997). *The psychic life of power: Theories in subjection*. Stanford: Stanford University Press.

DiGiuseppe, M. (2003). *Nelson biology 12*. Canada: Nelson Thomson Learning.

DiGiuseppe, M. (2004). *Nelson biology 11: College preparation*. Toronto: Nelson Thomson Learning.

Dimick, A. S. (2012). Student empowerment in an environmental science classroom: Toward a framework for social justice science education. *Science Education, 96*(6), 990–1012.

Dunlop, J. (2010). *McGraw-Hill ryerson biology 11*. Toronto: McGraw-Hill Ryerson.

Emdin, C. (2010). *Urban science education for the hip-hop generation*. Rotterdam: Sense Publishers.

Ethics. (n.d.). In Merriam-Webster's online dictionary. http://www.merriam-webster.com/dictionary/ethic.

Foucault, M. (1972). *The archaeology of knowledge*. New York: Pantheon Books.

Foucault, M. (1985). *The use of pleasure*. New York: Pantheon Books.

Foucault, M. (1986). *The care of the self (The history of sexuality, vol. 3)*. New York: Vintage Books.

Foucault, M. (2003). Ethics of a concerned self. In P. Rabinow & N. Rose (Eds.), *The essential Foucault, selections from essential works of Foucault, 1954–1984* (pp. 25–42). New York: New Press.

Foucault, M., & Senellart, M. (2010). *The birth of biopolitics: Lectures at the college de France, 1978–1979*. New York: Picador.

Foucault, M., Senellart, M., Ewald, F., & Fontana, A. (2007). *Security, territory, population: lectures at the Collège de France, 1977–78*. Basingstoke: Palgrave Macmillan.

Gerlach, N. (2011). *Becoming biosubjects: bodies, systems, technologies*. Toronto: University of Toronto Press.

Hodson, D. (2008). *Towards scientific literacy: A teachers' guide to the history, philosophy and sociology of science*. Sense Publishers.

Levinas, E. (1979). *Totality and infinity: an essay on exteriority*. The Hague: Martinus Nijhoff Publishers

Levinson, R. (2008). Promoting the role of the personal narrative in teaching controversial socio-scientific issues. *Science & Education, 17*(8–9), 855–871.

Mensah, F. M. (2009). Confronting assumptions, biases, and stereotypes in preservice teachers' conceptualizations of science teaching through the use of book club. *Journal of Research in Science Teaching, 46*(9), 1041–1066.

Merton, R. K. (1973). *The sociology of science: Theoretical and empirical investigations*. University of Chicago press.

Lemke, T. (2011). *Biopolitics: An advanced introduction*. New York: New York University Press.

Lock, R., & Ratcliffe, M. (1998). Learning about social and ethical applications of science.

Mills, S. (1997). *Discourse: The new critical idiom*. London and New York: Routledge.

Reiss, M. J. (1999). Teaching ethics in science. *Studies in Science Education, 34*(1), 115–140.

Reiss, M. (2008). The use of ethical frameworks by students following a new science course for 16–18 year-olds. *Science & Education, 17*(8–9), 889–902.

Rose, N. S. (2007). *Politics of life itself: Biomedicine, power, and subjectivity in the twenty-first century*. Princeton: Princeton University Press.

Sadler, Troy D., & Zeidler, Dana L. (2004). The morality of socioscientific issues: Construal and resolution of genetic engineering dilemmas. *Science education, 88*(1), 4–27.

Saunders, K. J., & Rennie, L. J. (2013). A pedagogical model for ethical inquiry into socioscientific issues in science. *Research in Science Education, 43*(1), 253–274.

Taylor, V. E., & Winquist, C. E. (2002). *Encyclopedia of postmodernism (Taylor & Francis e-Library ed.)*. London: Routledge.

Simons, M. (2006). Learning as investment: Notes on governmentality and biopolitics. *Educational Philosophy and Theory, 38*(4), 523–540.

Tolbert, S. (2015). "Because they want to teach you about their culture": Analyzing effective mentoring conversations between culturally responsible mentors and secondary science teachers of indigenous students in mainstream schools. *Journal of Research in Science Teaching*.

Van Rooy, W. (1993). Teaching controversial issues in the secondary school science classroom. *Research in Science Education, 23*(1), 317–326; *Education, 6*, 127–142.

Weber, M., & Andreski, S. (1983). *Max Weber on capitalism, bureaucracy, and religion: a selection of texts*. London: Allen & Unwin.

Weinstein, M. (2008). Finding science in the school body: Reflections on transgressing the boundaries of science education and the social studies of science. *Science Education, 92*(3), 389–403.

Zeidler, D. L., & Sadler, T. D. (2008). Social and ethical issues in science education: A prelude to action. *Science Education, 17*(8), 799–803.

Zeidler, D. L., Sadler, T. D., Simmons, M. L., & Howes, E. V. (2005). Beyond STS: A research-based framework for socioscientific issues education. *Science Education, 89*(3), 357–377.

Žižek, S. (2009). *First as tragedy, then as Farce*. London: Verso.

Chapter 4
Science Education and Subjectivity in (Bio)Political Context

> *"He doesn't know the sentence that has been passed on him?"*
> *"No," said the officer again, pausing a moment as if to let the*
> *explorer elaborate his question, and then said: "there would be*
> *no point in telling him. He'll learn it on his body."*
> —Franz Kafka, In the Penal Colony, (Kafka 2012, p.145)

Abstract In this chapter I argue that biopolitics can serve as an orienting concept for ethical and political engagement in science education. Using Hardt and Negri's (2000, 2009) notion of biopower and biopolitics, I argue that science education finds itself in the interstitial space between knowledges that govern and the apparatus of schooling. Science education is therefore a crucial site of resistance in (bio)political struggles against destructive forces of modern governance.

Keywords Biopolitics · Biopower · Governance · Science education · Foucault · Politics

In this chapter I want to further develop a biopolitical context for looking at subjectivity and ethics in science education. As I will argue, acknowledging that social struggles operate in a biopolitical context that pushes back on forces of biopower is not the same as engaging politically or ethically. Biopolitical struggles today require a participatory politics, which will look different depending on the topology of the struggles faced by educational communities. An engagement with ethics in science education that meets twenty-first century problems such as climate change and social inequality stands to gain by adopting an (ethical) commitment to political engagement. However, to theorize this in science education scholarship is difficult because the discipline has arguably never really been a place for theorization and creation, only problem solving and application. This needs to change. Science educators and scholars should have the option of, not just adopting theories, but creating and re-envisioning them for the goals of social and ecological justice [see Shulz's (2009) call to develop a field of 'philosophy of science education'].

© The Author(s) 2016
J. Bazzul, *Ethics and Science Education: How Subjectivity Matters*,
SpringerBriefs in Education, DOI 10.1007/978-3-319-39132-8_4

Biopolitics, as a way of understanding institutional relations, bodies, governing rationalities, and the use of objective knowledge, can help educators interrogate the relationship between knowledge and governance. Science is entangled in political agendas; it is sometimes both the cause of, and solution to, various social and ecological issues such as environmental destruction, racisms, heteronormativity, poverty, etc. (Harding 2006). Due to its entanglement with social and ecological issues, science education becomes a site, among many, where discourses of power work to order social life. Developing a biopolitical perspective requires that we relate pressing social and environmental issues as well as social theory in creative and interdisciplinary ways. This chapter will provide a context for engaging biopolitically in science education as a general way of thinking about the many contexts for ethical engagement today. I will delve into some of the theory behind a biopolitical perspective in an effort to explain why this perspective is useful for engaging many social and ethical issues that touch science education. I develop some of these ideas, such as the role of capitalism in biopolitics, in more detail elsewhere (Bazzul 2014).

As mentioned in the introduction, science education is located interstitially between governing knowledges (e.g. state curriculum and policy), and the objective, 'truth' knowledges of science. Thus, science education, in some ways more than other fields of education, carries the effects of power. It is therefore important to follow Harding's (2006) call for openness around social, cultural, and political issues in science and science education in order to nurture a relationship between "pro-democratic culture" and scientific knowledge (p. 51). As it stands, a pro-democratic relationship between scientific knowledge and the social world has competition from other forces. As Pierce (2013) argues, scientific literacy must take into account the current (bio)capitalist contexts now attempting to control science education for the benefit of a small minority of corporate elites. Pierce argues that reforms in education are bent on extracting '(bio)value' from students; and constituting an ethic, or subjectivity, of self-investment in those that have the means to make themselves into entrepreneurs. For Pierce, it is imperative that educators stand against 'corporate-minded' reforms that seek to commodify life by shaping students into human capital and self-investing subjects for the purposes of creating STEM surplus labour for multinational corporations, while putting forth a view that trees, water, plants, and animals exist primarily for business interests and the production of commodities in capitalist markets.

Biopower and Biopolitics

Biopolitics can be theorized in a variety of ways. The usage I intend is not just a kind of environmentalist interpretation where ecological concerns become the object of political interventions. Lemke's (2011) comprehensive introduction to biopolitics covers many different conceptions of biopolitics, including Agamben's (1998) concepts of *homo sacer* and *bare life*, and is a useful tool for understanding biopolitics in more detail. Continuing with our use of Michel Foucault, it is his conception of biopower and biopolitics that will be followed here. According to Lazzarato (2002), Foucault was "already pointing out in the seventies what, nowadays, is well on its way to being obvious: 'life' and living beings [le vivant] are at the heart of new political battles and new economic strategies" (p. 1). As mentioned in the last chapter, Foucault's work focused on the shift of governance centered on disciplinary practices of the body toward the management of populations (Foucault 1977, 1980). A shift from governing forces that impeded and punished, to forces focused on ordering forms of social life and making them grow (Foucault et al. 2007). No longer does governing rely on the power to end life, but rather on allowing some forms 'of living' to prosper and others to be stunted. These forms of governance are not mutually exclusive. Rather, it is more accurate to say that they represent two poles of governance that seek to control individuals at different levels, one acting directly on individuals, such as criminal punishment, and the other on large groups, such as school curriculum. (Lemke 2011). *Biopower* can be thought of as strategic forces of modern governance bent on controlling and extracting value from various 'forms' and levels of life. As was discussed in Chap. 3, this is achieved, in part, through the perspective that human beings are also biological beings. In this way, human 'life' is managed through discourses and practices of preventative health, statistical analysis, and infrastructure arrangements.

Political theorists Hardt and Negri (2009) emphasize the importance of subjectivity in today's social world, as capital invades all spaces of life where value is produced and extracted by mechanisms of biopower. Hardt and Negri (2000) call this broad phenomenon *Empire*—where the (re)production of subjectivities maintains the social, political, and economic order, which is becoming increasingly reliant on immaterial labour—the production of information and cultural content through media technologies. Empire is a global regime of biopower that regulates social life to a large extent through subjectification practices that create 'self-responsible' individuals meant to 'reactivate' this (bio)power in daily life (Rabinow and Rose 2006). Empire, as a globalized form of biopower, (re)produces subjects that in-turn reproduce our current socioeconomic system. Yet it is also in this context where struggles for democracy are waged and won biopolitically (Lewis 2007, p. 689).

Two Modernities

Hardt and Negri's (2000) vision of biopower and biopolitics, extending Foucault's conceptualization and borrowing from Deleuze and Guattari's (1988) focus on imminence, involves two countervailing forces of modernity: *constituent* powers and *constituted* powers. Constituent powers refer to the immanence and creativity of human capacities—while constituted powers involve appeals to transcendent authority and work to establish hierarchical order. Hardt and Negri (2000) claim that modernity's defining positive feature is the privileging of immanent creative forces, where human beings realize the potential for creative, free thought that exists here and now. In sixteenth century Europe[1] knowledge begins to shift away from a reliance on transcendent sources toward practices of thinking that 'dares to know' (Foucault 1984). As Hardt and Negri (2009) put it, the powers of the heavens now begin to reside on earth. It is in the immanent force of modernity where the freedom to become resides. However, transcendent forces of modernity arose simultaneously to establish hierarchical dominance over these forces, by bringing all differences, all multiplicities, under the control of universalities and disciplinary apparatuses, such as the *rule* of private property.

In science, reference to fundamentalist or transcendent views of biology, for example, could involve and appeal to a radical realist view of evolutionary psychology, where alternative sexualities are labeled as aberrant because they do not fit some static, transcendent version of human nature. Understanding modernity as countervailing two forces is useful to an understanding of biopower and biopolitics; with constitutive transcendent forces being associated with biopower and the struggle to nurture and free constituent, immanent forces associated with biopolitical struggles. If biopower encompasses the structures, forces, discourses and practices bent on extracting value from life, engaging biopolitically involves a critical response to biopower that promotes the imminent power of cooperation and creativity inherent in community-oriented, democratic forms of life. Another way of saying this is that biopower attempts to dominate and control social life from above in hierarchical, territorializing forms, while biopolitics is the response to these forces from below. To escape ecological catastrophe and social inequality, we must gravitate towards the imminent forces of modernity over ones that seek to control, exploit, and destroy differences. Working against these controlling forces in education involves asking why we take for granted certain ways of thinking about humans through science, or why we find these ways of thinking natural.

Rabinow and Rose (2006) define biopolitics as intentions "[t]o embrace all the specific strategies and contestations over problemitizations of collective human vitality, morbidity and mortality; over the forms of knowledge, regimes of authority and practices of intervention that are desirable, legitimate, and efficacious" (p. 197). They contend that analyses should focus on how biopower operates on multiple, small-scale fronts, from 'what is best for us', to what kinds of warnings or markers citizens, consumers, and workers should pay attention to regularly. We can think of these analyses as taking place on both *macro* and *micro* poles; that is, analyses that involve both the practices of daily life and large-scale policy, law, and institutions (Bazzul 2016). For science educators it is at the micro level—how specific education policy is constructed or what happens in classrooms and community interactions—where we can be most effective in engaging the focus of biopower.

Science education specifically deals with discourses that carry the effects of biopower; discourses such as biotechnology, biology, toxicology, and food science

---

[1]It is certainly debatable whether the immanent character of human thought made the most progress in Europe. However, if nothing else, Hardt and Negri's argument is that European modernity is both a source of a wealthy tradition and the rise of a great and often terrible power.

education. Specific elements of biopower can be traced and examined in science education, according to the following categories:

- Authorities/experts that speak objectively about the 'character' of human life, which may include: genetic counselors, scientists, educational authorities, medical experts, etc.
- Strategies to manage collective existence in the name of life or health. These strategies constitute relationships between race, sex/gender, ethnicity, biological or genetic citizenship.
- Practices of subjectification through which students are brought to work on themselves under the aforementioned authorities in the name of the 'character' of human life.

These aspects of governance can be traced to specific historical contingencies linked to official economic and scientific knowledges. Although biopower involves practices of subjectification, the vast majority of these practices are not obvious because they inhabit micro social spaces. However, it is at these micro-locations where social struggles become biopolitical in terms of pushback and a reworking of biopower. Getting back to specifics, science education may exercise biopower when:

- Knowledge is applied to human life, through designated authorities.
- It is involved in health and population management or problems of collective existence.
- 'Who' can think and act 'scientifically' is delineated in discourses and practices (Chap. 2).

The question is not whether science education exerts the forces of biopower, but how much, and how do science education communities push back biopolitically?

Hardt (2010) gives some direction in regard to education that works against the forces of Empire. According to Hardt, education must work to tear down hierarchies that keep oppressive social orders in place. Although there is inherently power in the way educational relationships are formed, tearing down hierarchies can proceed as the educational process unfolds. As stated in Chap. 3, education must de-privilege the voices of experts who tell educational communities they have no control or choice. Power and wealth come from below, not through efficiencies, or the privileging of profits at the expense of development for social and ecological justice. Education needs to protect those on the edges of social exclusion and economic production, because they have the greatest capacity to push back against unjust systems (Empire) that do not/cannot function for them. Education must also become oriented toward protecting the commons, putting the natural and social commons to work for the service of all life (Means 2013).

Biopolitics and Science Education

Taking a biopolitical approach in science education means that rather than looking to simply blame commercial interests or nefarious government agendas for social and ecological problems, we come to see how biopower operates at the level of categorization, normalization, and subjectification—for example how life is cast—ethical, healthy, sick, responsible, etc. Further, what are the politics of 'self' in an age of genomics? The question of subjectification is made especially complex by the many networks made available through communications technology (Deleuze 1992). How does biopolitics give context to a variety of social struggles in science education? Let us take a look at some examples from previous chapters: racisms/colonialisms; neoliberalisms; struggles of sex/gender and sexuality; as well as biotechnologies and their influence on the constitution of subjectivity.

Colonization and the Power to Make Die

Science students can experience conflict between their multi-dimensional, racialized, sexed, subjectivities and the assumed universal, white, male, heterosexual, subject of science (Irigaray and Bové 1987; McKinley 2008). Colonialisms and racisms are an example of the controlling forces of modernity, and can be viewed as a biopolitical problem (Stoler 1995), which involves the development of science. As Stoler argues, racism is an incessant social war driven by technologies of purification, and is internal to the Modern Western State. Colonization solves one problem of the transition from disciplinary to a regulatory society—the management of populations discussed in Chap. 3—where the 'right to kill' (discipline of individual bodies) is at odds with the goal of fostering and controlling life (at the level of populations). Namely, it allows a state to simultaneously maintain brutal regimes of bodily discipline where colonizing discourses designate 'who' must die for the sake of those who can live, and in this way "establishes a positive relation between the right to kill and the assurance of life" (Stoler 1995, p. 84). The logic being: *the more subaltern peoples die, the more privileged subjects live*. Colonization retains and uses disciplinary (bio)power, as an always-incomplete cleansing of the social body (Lemke 2011). Scientific cultures may exercise epistemological and ontological forms of the 'right to kill' when they have a role in determining 'what knowledges must die (indigenous/local) for others (scientific) to live'? Other critical questions may include: Who is 'othered' through discourses and practices of science education? How is the health of one person in direct relation to the poverty/ill-health or death of another?

Neoliberalisms

Reforms in science education have followed a neoliberal agenda, which gives primacy to global capitalist interests over those of local communities (Bencze and Carter 2011; Tobin 2011; Weinstein 2012). Neoliberal subjectivities are an important part of biopower today as state-corporate partnerships now frame schooling as a place where self-investing, entrepreneurial subjects are produced. Promotion of the entrepreneurial self in science education is becoming more widespread, for example with the celebration of private entrepreneurs like Craig Venter and in the way students are led to invest in their own human capital through discourses around career choices. How do education discourses constitute a subject's relation to private property, corporate interests, and their connection to scientific research? (Bazzul 2012).

The Regulatory and Disciplinary Poles of Sex/Gender and Sexuality

Foucault (1980) positions sexuality between the two modes of power/governance; the disciplinary/bodily and population/regulatory. The power effects that help shape performances of sex/gender and sexuality, are both situated on the body and controlled or normalized at the level of populations. Norms and regulations regarding hygiene, health, and relationships make sex/gender and sexuality a biopolitical issue par excellence! For example, life science textbooks' discourses express concerns over the growth of aids in a population and prescribe particular individual (preventative) sexual behaviors. The AIDS virus is therefore both a biological and social phenomenon that exposes how biopower works to distribute the effects of power both at the level of the population and the individual. Pushing back against this power, say in re-working and questioning how LGBTQI people are positioned by these discourses, is one way educational communities can engage biopolitically.

As Anne Fausto-Sterling demonstrates, the constitution of sex/gendered and sexualized subjectivities in science works to (re)produce heteronormative, male-gendered science research (Fausto-Sterling 2012). In terms of subjectivity, philosopher Luce Irigaray (Irigaray and Bové 1987) challenges science's 'male voice' (add white and straight!) under its guise of universality. Irigaray asks the following critical questions in relation to science:

> Does the alternative become either do science or "be a militant"? Or again to continue to do science and to divide yourself up into different functions, several persons or characters? Should the truth of science and that of life remain separate, at least for the majority of researchers? What science and what life is then under discussion? Especially since life in our time is greatly dominated by science and its techniques (p. 78).

The relation between sex/sexuality and gender and science and science education will continue to constitute various subjectivities and forms of life, making it an important site for biopolitical engagement.

Biotechnology and Biosubjects

According to Lazzarato (2002), genomic research, artificial intelligence, and biotechnology are new frontiers for the exercising of biopower. For example, the privatization of seeds (plant life) in the global 'south' involves social policies intimately related to practices of science research—demonstrating that biotechnologies are never politically neutral (Carter 2011). Shiva (Shiva and Moser 1995) asserts that "not until diversity is made the logic of production can diversity be conserved" (p. 207). In biopolitical terms, this means challenging controlling forces of modernity such as the rule of private property, where some (White, Male, Global North) bodies control the material circumstances of others.

Biotechnology is changing how human beings are governed by changing the very limits of the body. New forms of subjectivity stemming from biotechnologies increasingly trouble dualisms of natural/artificial, human/animal, and present/future (Haraway 1997). This subject is increasingly dependent upon a fragmented body, increasingly outside of humanist ethics, yet firmly within capitalist relations and in conversation with the entrepreneurial subject and the subject under surveillance (Gerlach 2011, pp. 6–7). Bodies today must also bare the 'biological gaze' at the molecular level (Keller 1996), where they can be "informed, sold, killed, manipulated, reproduced, copied, and circulating along networks of exchange and knowledge production" (Gerlach et al. 2011, p. 9). These changes to everyday identities, enforced by authorities such as scientists, informational media, counselors, and education systems, will also lead to new forms of political activism surrounding science related issues.

A Biopolitical Context as an Ethico-Political Frame

> Let us begin with indignation, then, as the raw material for revolt and rebellion. In indignation, as Spinoza reminds us, we discover our power to act against oppression and challenge the causes of our collective suffering. In the expression of indignation our very existence rebels.
>
> —Hardt and Negri (2009, p. 236)

Viewing science education as a site for biopolitical engagement can help students and educators engage issues of inequality and environmental destruction by working within networks and pushing back against institutions to realize new subjectivities and forms of life. Engaging biopolitically means moving away from power *over* life toward the power *of* life in order to re-work how we come to see ourselves and the world. If ethics in science education is to engage (bio)politically, it must disrupt biopower, which orders and extracts value from populations, and re-work this power towards collective justice. It will also involve creating the freedom for students to think differently, and see themselves differently, in relation to science related social and environmental issues. Education for emancipation, from a biopolitical perspective, cannot be conducted without engagement in science education, because scientific disciplines (bolstered by the power of states) have institutional authority to 'speak truth' about what forms of social life (and biological!) are socially valuable.

Practices of subjectification, which bring students to embrace certain modes of conduct, should be the focus of educational research that works toward transformative social change. Indignation stemming from the collective oppression of the marginalized and destruction of the biosphere is a necessary first step to help locate what needs changing. According to Pierce (2013), science educators must promote scientific literacies that "link learning and teaching science to practices and social movements that are actively resistant to biocapitalist visions of the future, ones that represent cultural practices rooted in communities producing biodemocratic life" (p. 108). Critical analyses of subjectification practices in science education are a crucial 'second step' in making these literacies a reality. As stated above, science education is, perhaps, the best place to engage biopolitically as it is situated within multiple authoritative, normalizing forces, practices, and discourses that address the everyday lives of both individual subjects (students) and the population at large.

References

Agamben, G. (1998). *Homo sacer*. Stanford: Stanford University Press.

Bazzul, J. (2012). Neoliberal ideology, global capitalism, and science education: Engaging the question of subjectivity. *Cultural Studies of Science Education, 7*(4), 1001–1020. doi:10.1007/s11422-012-9413-3.

Bazzul, J. (2014). Science education as a site for biopolitical engagement and the reworking of subjectivities: Theoretical considerations and possibilities for research. In *Activist Science and Technology Education* (pp. 37–53). The Netherlands: Springer.

Bazzul, J. (2016). Biopolitics and the 'subject' of labour in science education. *Cultural Studies of Science Education,* 1–14. (online first).

Bencze, J. L., & Carter, L. (2011). Globalizing students acting for the common good. *Journal of Research in Science Teaching, 48*(6), 648–669.

Carter, L. (2011). Gathering in threads in the insensible global world: The wicked problem of globalization and science education. *Cultural Studies of Science Education, 6,* 1–12.

Deleuze, G. (1992). *Post script to societies of control*, vol. 59, pp. 3–7.

Deleuze, G., & Guattari, F. (1988). *A thousand plateaus: Capitalism and schizophrenia*. Bloomsbury Publishing.

Fausto-Sterling, A. (2012). *Sex/gender: Biology in a social world*. New York: Routledge.

Foucault, M. (1977). *Discipline and punish: The birth of the prison*. New York: Pantheon Books.

Foucault, M. (1980). *The history of sexuality. Vol. 1: An introduction*. New York: Vintage Books.

Foucault, M. (1984). What is enlightenment? In P. Rabinow (Ed.), *The Foucault reader* (pp. 32–50). New York: Pantheon.

Foucault, M., Senellart, M., Ewald, F., & Fontana, A. (2007). *Security, territory, population: lectures at the Collège de France, 1977–78*. Basingstoke: Palgrave Macmillan.

Gerlach, N. (2011). *Becoming biosubjects: bodies, systems, technologies*. Toronto: University of Toronto Press.

Haraway, D. (1997). *Modest-witness@second-millenium. FemaleMan-meets-oncomouse: Feminism and technoscience*. New York: Routledge.

Harding, S. (2006). *Science and social inequality: Feminist and postcolonial issues*. Urbana: University of Illinois Press.

Hardt, M. (2010). The militancy of theory. *The South Atlantic Quarterly, 110*(1), 19–35.

Hardt, M., & Negri, A. (2000). *Empire*. Cambridge: Harvard University Press.

Hardt, M., & Negri, A. (2009). *Commonwealth*. Cambridge: Belknap Press of Harvard University Press.

Irigaray, L., & Bové, C. M. (1987). Le Sujet de la Science Est-ll Sexué?/Is the Subject of Science Sexed?. *Hypatia*, 65–87.

Kafka, F. (2012). *The complete stories*. New York: Schocken Publishers.

Keller, E. F. (1996). The biological gaze. In G. Robertson, M. Mash, L. Tickner, J. Bird, B. Curtis, & T. Putnam (Eds.), *Future natural: Nature, science, culture* (pp. 107–121). London: Routledge.

Lazzarato, M. (2002). From biopower to biopolitics. *Pli: The Warwick Journal of Philosophy, 13*(8), 1–6.

Lemke, T. (2011). *Biopolitics: An advanced introduction*. New York: New York University Press.

Lewis, T. (2007). Biopolitical Utopianism in educational Theory. *Educational Philosophy and Theory, 39*(7). doi:10.1111/j.1469-5812.2007.00316.x.

McKinley, E. (2008). From object to subject: Hybrid identities of indigenous women in science. *Cultural Studies of Science Education, 3*, 959–975.

Means, A. (2013). Creativity and the biopolitical commons in secondary and higher education. *Policy Futures in Education, 11*(1), 47–58.

Pierce, C. (2013). *Education in the age of biocapitalism: Optimizing Educational Life for a Flat World*. New York: Palgrave Macmillan.

Rabinow, P., & Rose, N. (2006). Biopower today. *Biosocieties, 1*, 195–217.

Schulz, R. M. (2009). Reforming science education: Part I. The search for a philosophy of science education. *Science & Education, 18*(3–4), 225–249.

Shiva, V., & Moser, I. (1995). *Biopolitics: A feminist and ecological reader on biotechnology*. London: Zed Books.

Stoler, A. L. (1995). *Race and the education of desire: Foucault's History of sexuality and the colonial order of things*. Durham: Duke University Press.

Tobin, K. (2011). Global reproduction and transformation of science education. *Cultural Studies of Science Education, 6*, 127–142.

Weinstein, M. (2012). Science, science education and the politics of neoliberal exceptionality. In J. R. McGinnis, S. J. Lynch, W. C. Kyle, & T. A. Sondergeld (Eds.), Re-imagining research in 21st century science education for a diverse global community proceedings of the 2012 National Association of Research in Science Teaching international conference. Indianapolis: NARST e-Publications (CD-ROM).

Chapter 5
Egalitarian Politics and the Dimensions of an Ethical Self

> *There is no specific moral action that does not refer to a unified moral conduct; no moral conduct that does not call for the forming of oneself as an ethical subject; and no forming of the ethical subject without modes of subjectivation and an ascetics or practices of self that support them. Moral action is indissociable from these forms of self-activity, and they do not differ any less from one morality to another than do the systems of values, rules, and interdictions.*
>
> —Foucault (1985, p. 28)

Abstract With the recent move towards activism and engagement with ethical issues through science education it is more important than ever to examine the ethical dimensions of educational policy and practice. In this chapter, I take a theoretical approach to ethics by laying out Foucault's (1985) ethical relations of self. I argue that politicizing these ethical relations of self is the most viable way to meet the challenges of growing social inequality and climate change. While ethics and egalitarian politics may seem to go hand in hand, I argue that there are some antagonistic elements to the merging of ethics and politics—and suggest some ways to successfully merge emancipatory political perspectives with a focus on ethics as they relate to relations of self. While science education has not been a space where scholars have been encouraged to theorize, this must change if educators are going to find new ways of combatting social and environmental problems.

Keywords Ethics · Subjectivity · Self · Foucault · Materialisms · Ontology · Science education · Critical theory

Science educators must pay closer attention to 'the subject' of ethics in the twenty-first century. Social inequality, environmental catastrophe, the rise of new urban slums, and exclusions from global citizenship are destroying and depleting the social and natural commons. Educational communities need to ethically respond to these challenges in order to re-evaluate how we approach our shared world. The constraints

© The Author(s) 2016   53
J. Bazzul, *Ethics and Science Education: How Subjectivity Matters*,
SpringerBriefs in Education, DOI 10.1007/978-3-319-39132-8_5

to ethical ways of being today mostly have to do with the hegemony of a global elite (Global North, White, Wealthy (1 %), Male, Western, Heterosexual, Able-bodied), and necessitate a renewed engagement with ethics toward more just forms of existence. Current forms of global domination that exploit the natural commons and prevent just social relations, what Carter (2011) deems *wicked* problems, cannot be allowed to stand unchecked—as Jameson (1991) put it, we must challenge blind adherence to the 'almighty' market. These various forms of global hegemony, sometimes hiding under the universalisms of science (Harding 1993), along with the dismantling of public education by private interests (Lipman 2013), is an ethico-political test for education, including science education. How will science learning communities help inculcate the necessary ethical outlooks and subjectivities needed to meet these wicked problems? What forms of ethical and political engagement challenge oppressive social relations?

Different modes of engagement will be necessary if science education is to fulfill a much needed sociopolitical role (Roth and Barton 2004). Pedagogies that position students as 'typical citizens', enacting typical citizen actions such as voting, etc., may not be sufficient to meet twenty-first century challenges (Bazzul 2015). This chapter expands on how ethical actions/decision making must be considered along three dimensions of relations of self (see Chap. 3): self to self; self to other; and self to the world. A focus on the constituted nature of subjectivity is key in the development of an ethical subjectivity, and the creation of the contexts for different politico-ethical forms of action.

Ethical Subjectivity and Relations of Self

Educational practices and discourses do not shape pure rational subjects, or blank slates, but actually work to constitute the very subject positions students and educators recognize as 'natural' or 'free' (Butler 1997; Foucault 1982; Bazzul 2014a). This happens in school science, including when the goal is to produce scientifically literate, engaged citizens as Hodson (2003) advocates. Discourses of science education delineate how ethical issues become visible and how students are to engage ethically. As we saw in Chap. 3, the options for students to engage ethically largely involve suggesting changes to policy as well as changes to their personal lifestyle (see also Bazzul 2014b). Science education discourses afford certain modes of ethical engagement and constrain others, and in this way constitute 'ethical subjectivities' in relation to science. Not only do students come to see some issues, and modes of engagement, as natural and commonsensical, they also come to see themselves, their identities, as ethical actors in particular ways. Engaging ethics in science education requires looking at how 'ethical subjectivities' are constituted through educational discourses, as well as viewing the classroom as a site where crucial ethical actors are formed. Otherwise, ethics will cease to be a useful concept for transformative science education educators in the age of what Hardt and Negri (2000) call *Empire* (see Chap. 4). In this chapter I argue that the relations of self that comprise an ethical subject, while informed by many contexts, codes, emotions, etc., should take on an emancipatory political tone if science education is going to meet twenty-first century problems such as climate change and social inequality.

To do this, I draw from Foucault's (1985, 1986) *History of Sexuality* volumes two and three, which develop the idea of relations of self, or care of self, in Greco-Roman antiquity. One of the key questions Foucault asked, before Christianity and modern technologies of power arose in Europe, was how did someone come to see themselves as 'being' ethical? While this may seem irrelevant to education scholars in the twenty-first century, to get a better view of our present approach to being ethical, it is sometimes necessary to visit different contexts (Foucault 1985). Foucault's analyses can shed light on what it may mean to be a 'subject of ethics' today. In the History of Sexuality Volumes 2 and 3, Foucault demonstrates that 'being ethical' cannot be reduced to a code, the treatment of others, or the kind of person one wants to be (Foucault 1985, 1986). Rather, it is a confluence of these aspects that must be understood against a set of relations of self.

Foucault's work can guide how educational communities engage issues of ethical importance related to science by examining how curriculum and pedagogy constitute students as ethical subjects along three interrelated axes: self's relation to self; self's relation to others; self's relation to the world. Figure 5.1 outlines these axes.

Foucault (1985) argues that ethical codes, religious, commercial, civic, sexual, etc., cannot alone dictate, predict or comprehensively advise how a person is to behave ethically, but are just one dimension of this question. Instead, the three

Michel Foucault's Ethical Relations of Self

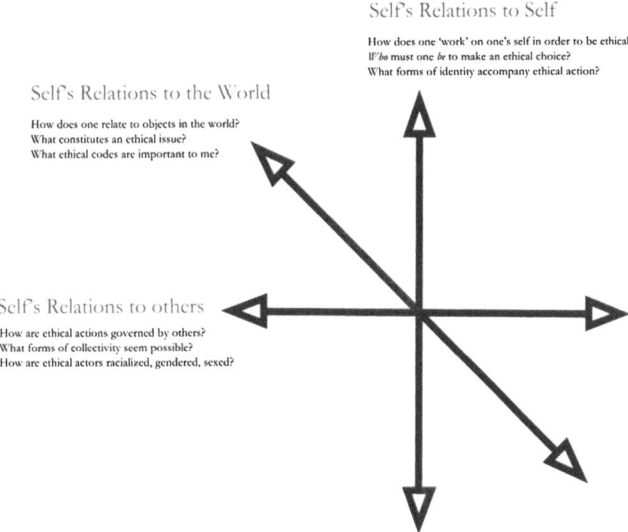

Fig. 5.1 Three axes of the ethical self as presented in Foucault's (1985) *use of pleasure*

Table 5.1 Different approaches to ethics based on relations of self to self, others, world

Relation	Individualistic	Philanthropic	Politicized
to Self	Ethical actions are a personal matter, and always come down to personal choice	I see myself as having some responsibility to give back to my community	My identity is collective, and my efforts are dedicated to collective justice
to Other	Change happens only when others take personal responsibility for their actions	Everyone should give back to the communities where they live and work	Collective justice requires activism at the level of institutions and everyday life
to World	Humans seek their own advantage, and ethical actions align with a person's interests	The world is a community of individuals who should give of themselves	Extinction, suffering, and wellbeing, are tied to inequality and collective existence

simultaneous relationships of self, shown in Fig. 5.1, must be taken together as coordinating dimensions affecting how a subject acts/behaves/thinks ethically. In other words, ethical action does not consist of a set of preconceived tenets, but is enacted in relation to an ethical actors' relation to self, others, and the world—this enactment can be seen as part of an 'ethical subjectivity' and/or being an 'ethical subject'.

As discussed in Chap. 3, educational scholars need to view ethics in broad terms, *what one ought or ought not to do,* to operationalize it as a concept for research. This is because ethics can entail any sort of practice related to 'a right way' of being, from dietics, to business, to a friendship with a transcendental being. Education is a site where we can interrogate ethical practices and transform the very basis for ethical action. Following Foucault, I maintain that an analysis of ethical subjectivity, and/or taking ethical action, involves considering how students view their relation to self, others, and the world(s), as well as instilling a politically active approach to science through education.

When students and teachers engage issues of ethical importance in relation to science, they often ask a basic ethical question: '*What should we do*'? They may find some ethical actions/positions obvious, such as: *gather and present data!*; *recycle!*; *spend less!*; *donate!*; *become more efficient!*; *change the law*!. However, ethical actions can mean different things, are enacted for different reasons, and have different outcomes, depending on how students and teachers see themselves in relation to the world, others, and themselves. Table 5.1 describes how a student could possibly hold different relations of self, and how these might affect ethical action, from *individualistic*, *philanthropic*, and *politicized* perspectives. What I propose is that a politicized form of ethical engagement is more conducive to meeting twenty-first century ethical issues. Although the categories below are abstractions, and not mutually exclusive, they illustrate that while everyone has the potential to be 'ethical', the grounds and context for doing so vary greatly.

When an individual or educational community decides to take ethical action they can do so under entirely different understandings of self. For example, people can engage in environmental cleanup within an ethical frame of 'personal accountability' (Individualistic); 'dutiful community participation' (philanthropic); or 'collective activism' (politicized). Thinking of these ethical frames as abstractions allows educators to ask after the relations of self that are at play when making ethical decisions regarding issues related to science. A hypothetical example of students holding different relations to self, individualistic, philanthropic, and politicized, might unfold like this:

Student A: Individualistic Relations of Self

to Self: I engage in community cleanup because I am responsible and need service hours.

to Other: I engage in community cleanup to demonstrate acceptable behaviour to others.

to World: For the world to be cleaner, everyone needs to take responsibility for cleanliness.

Student B: Philanthropic Relations of Self

to Self: I am a member of the community and my cleanly behaviour reflects this.

to Other: I give my time to clean an area for those that do/can not do it themselves.

to World: The world is comprised of communities that require assistance from people like me.

Student C: Politicized Relations of Self

to Self: My work fits within a larger structural struggle of environmental degradation.

to Other: Privilege divides those who have and do not have control over where they live, and whether or not those areas are or are not clean environments.

to World: Relations of power govern how humans and other living things can exist in the world.

I contend that discourses in science education can be analyzed in terms of whether they enable particular ethical actions according to particular relations of self. That is, looking to see how relations of self operate as a framework that contextualizes ethical actions. These considerations may help us abandon forms of ethical engagement that maintain and reproduce possessive individualism and ineffectual forms of ethical engagement that barely address important environmental or social issues. And why might educators want to abandon these forms? Simply put, status quo citizenship, and the tepid response to climate change and inequality by governments, does not seem to be making much of an impact. I argue that educational communities need to 'politicize' ethical relations of self, where politics begins to inform how we see ourselves, others, and the world in relation to ethical issues related to science. However, from a theoretical perspective, uniting politics and ethics is not as simple as it may sound.

In the final two sections I will talk about some of the theoretical problems of bringing ethics together with politics. Essentially this has to do with conflicts of equality and the right to speak. If, under democracy, every person has the equal right to speak, be heard, and decide, how do we solve the problem that some people, sometimes, have a *better ethos* of thought and action. Or to put it another way, how do we reconcile the fact that sometimes certain people need to be heard more than others, within the logic of equality. While this may not seem like a problem worthy of consideration, bringing about social and ecological justice through ethical action may require educators and scholars to think about why it may be so difficult to make school and society politically progressive from the perspective of democratic equality.

Uniting Ethics and Politics: Equality and the Right to Speak

Since many ethical issues related to science implicitly involve politics, it is important for science educators to nurture the political consciousness necessary for a critical citizenry. So far, I have argued that approaches to ethics need to be politically oriented, and involve politicized relations of self, to contend with social inequality and climate change. Defining politics is difficult because it can mean many different things. Politics sometimes has unsavoury connotations, and is often understood as the opposite to truth. I take philosopher Rancière's (Rancière and Corcoran 2010) view of democratic politics, which involves interventions, based on dissensus and disagreement, into the taken-for-granted (sensible) order of things in the name of equality.

According to Rancière, *political subjectivity* refers to an outlook where one endeavours to change the status quo, its assumptions and rational grounds, in the name of equal inclusion (Rancière and Corcoran 2010). What I am proposing is that an ethical subjectivity needs to establish this outlook within the field of relations of self that guide ethical behaviour. In this way, ethics can become co-extensive with democratic political engagement, rather than something based on individualism or token charity. Foucault's analyses help lay the groundwork for a re-examination of ethics as critical praxis. Ethics, as a set of relations, can become a mode of political participation that also acts as a practice of self-resistance against technologies of (bio)power that govern our *conduct*—for example, through discourses of neoliberal governmentality, whereby individuals are brought to invest in themselves (Foucault and Senellart 2010).

Neoliberal education reform shapes individual and organizational conduct for the purposes of human capital production through entrepreneurialism and private invest-ment instead of the needs of communities (see Brown 2015). It is important to consider how conduct is governed in modern times, because it is this conduct that is central to an *'ethos' of living* in terms of how it adds legitimacy to actions. In other words, our conduct is (bio)politically situated, and this (legitimized) conduct implic-itly carries 'ethical content'. Neoliberal capitalism forms an ethos, or a particular way of acting, by shaping the way people relate to each other, the social, ecological contexts of our lives, as well as how people see themselves. In this way ethics can be *completely* in line with oppressive forces—even when practiced with the best intentions! This is one of the reasons why Jacques Rancière is wary of a 'return to ethics' over politics (Rancière 2006). A depoliticized ethics is not a form of resistance to consumerism, ecological destruction, and social inequality in neoliberal capitalist times. As political theorist Lazzarato (2013) argues, practices (care) of self, crafting a worthwhile life, certainly do have a neoliberal capitalist 'version'.

And just as capitalist relations allow for, and even require, an ethical subjectivity and relations of self—engaging in acts of political resistance to these relations also requires a set of relations of self. That is, to recognize one's self as a legitimate ethical speaker/actor—an 'ethos' involving relations of self to self, self to other, and self to world. Lazzarato (2013) summarizes Foucault's conceptualization of ethos as a matrix where experience, behaviour, and existence are articulated together for subjects. The 'big' question I am trying to get at here is: what kind of ethos and relations of self are required for democratic engagement in social and ethical issues related to science? This question is more challenging than it may appear, and my solution thus far has been to politicize these relations of self in/through science education.

The Contradiction of Ethics and Politics

On a general level, one of the problems of instituting democratic practices, does not reside in a realization of the notion of political equality, but rather ethical differentiation, or how a political community determines *good* from *bad*. This is because they do not function with the same principles (Foucault 1985). For Lazzarato, egalitarian politics, based on the principle of equality, is not sufficient to the critique of neoliberalism or neoliberal liberty because it ignores the fact that neoliberalism entails a (political) desire to re-establish hierarchies based on legitimate critiques of egalitarian twentieth century socialism. This is to say that if politics means the establishment of equality, there still needs to be space in the political arena for what is 'good' or 'bad' about how equality is enacted—in short *politics needs ethics*. In a radical democracy, all are equal, but an ethos, or several, are needed by which to operate in the world.

I maintain that a politicized 'ethos' of equality, manifested in relations of self to self, world, and others, would inform ethical stances against social inequality and ecological collapse by opening new possibilities, an ethics, of collective action. Teachers, students, (future citizens) would come to recognize an ethos of equality in other people, that is, the ability or opportunity to speak about what is ecologically and socially just. They would also not recognize a legitimate ethos as belonging to individualistic modes of life. Such an ethos, one that is politically active in the sense of Freire's (2000) conscientização, risks delegitimization from those who will not recognize this ethos as legitimate and/or afford its adherents the equal right to speak and act. Indeed, the challenge for political communities fighting social inequality is to afford each other the **equal** right to speak! An ethos of equality must form the basis of political action toward social and ecological justice. To meet the challenges of climate change and social inequality, politicized democratic principles should form the ethical relations of self, and seek to afford the 'equal right' of other beings to be, act, speak, and think.

So. Politics cannot be enacted without an ethos, or a variety of them, that encompass relations of self because subjects operate within different contexts, discursive fields, where an ethos, a position of speaking/acting or being, must take shape. If ethics is to engage the major social and environmental issues of today it must infuse political principles of equality along the three axes of relations of self. This is because political principles are meaningless without modes of self-conduct, self-examination, and work-on-the-self that enact them. Politics requires a complementary ethics, much the same way that consumerism requires its own ethics. Freire's (1972) lifelong praxis is a good example of an ethical self cast alongside the politics of equality, struggle and transformation, connecting a political subjectivity with an ethical self—a differentiation in values.

A Politicized Ethics of Equality and Resistance!

Subjects emerge from vast assemblages of material and discursive elements that cannot be fully known, but the various ways we rework our relations to ourselves, to others, and to the world also comprise part of this subjectivity. As Lazzarato (2013) notes of life under late capitalism, it is impossible to separate ethics from economy and politics. Paying attention to relations of self can disrupt the control of conduct through biopower and consumerist ideologies. Žižek (2011) outlines the ironic example of Starbucks'™ *ethos*® water where consumers are buying a feeling, freedom from guilt enacted upon ethical relations of self somewhere in between the philanthropic and individualist categories suggested above ("I am what I buy"! "I am part of the solution, I care"). Green consumerism is simply enacting an individualistic ethos along a particular (capitalist) set of relations of self.

Experimentation in modes of 'being', or developing an ethics, is fundamental to (science) education for transformative change. Foucault (1985, 1986) maintains that when a subject enacts relations of self to self, other, and the world, they become a political subject, developing the capacity to resist the social order. According to Foucault, it is when a subject resists, or faces power, that life is most visceral (Deleuze 1988). In this way, when an ethical subject becomes politicized, taking on the structural causes of inequality, they will experience intense forms of joy and pain through opposition. It is important to keep asking questions about the relationship between self and ethics. How do educators and students remake themselves in a particular time and place into ethical subjects? What modes of resistance and practices are available to us? As Deleuze (1988) puts it: "What can I do? What do I know? what am I?". Or, to paraphrase him: do not the shifts to hyper-capitalism find an unexpected 'encounter' in the slow emergence of a new (ethical) self as a form of resistance (Deleuze 1988, p. 115)?

The overall goal of this research brief was to propose a *politico-ethico* approach to education and science education that, on a literal level, connects ethics with politics. I have yet to trace what ethical relations of self are active/activated in curricula or classroom settings. Make no mistake, currently there is already a particular set of relations of self that govern how students and teachers approach ethics (see Chap. 3). I contend that these relations of self are seldom politicized, but are instead individualistic or, at best, philanthropic. If teachers and students are going to engage the wicked problems of today they must politicize science education. These politics must be informed by biopolitical contexts, engage in forms of dissensus, and employ methods of assembling the natural and social commons.

Conclusion: Leaving an Anthropocentric Ethics Behind

I have argued that science education discourses work to constitute particular ethical subjectivities by affording and limiting choices for students to engage ethically. From the textbook analysis presented here, it seems that very few choices for collective action are offered to students. I maintain that politicizing ethics, and considering relations of self to self, other and the world is one form of ethical engagement that can better meet twenty-first century problems such as social inequality and climate change.

However, this discussion of ethics has been very human-centric. The future will also require the consideration of non-human, even non-living, entities as worthy of equal consideration and protection. Indeed the twenty-first century will have to shed its anthropocentrism to lesson the destructive effects of the Anthropocene. Such a move will require reconsidering the fundamental assumptions of our ontological and epistemological understandings of human beings, phenomena, and the responsibility we have to act in a connected world. Barad's (2007) work on reshaping what it means for phenomena and individuals to emerge in an intra-active world will be pioneering in this endeavour. Although, because the educational community is human, an anthropocentric ethics will be needed; these ethics will need to be seen against a backdrop of larger concepts of life, the natural and social commons, what all entities on planet earth share together.

References

Barad, K. (2007). *Meeting the universe halfway: Quantum physics and the entanglement of matter and meaning*. Duke University Press.

Bazzul, J. (2014a). Critical discourse analysis and science education texts: Employing Foucauldian notions of discourse and subjectivity. *Review of Education, Pedagogy, and Cultural Studies, 36*(5), 422–437.

Bazzul, J. (2014b). Tracing "ethical subjectivities" in science education: How biology textbooks can frame ethico-political choices for students. *Research in Science Education, 45*(1), 23–40.

Bazzul, J. (2015). Towards a politicized notion of citizenship for science education: Engaging the social through dissensus. *Canadian Journal of Science, Mathematics and Technology Education, 15*(3), 221–233.

Brown, W. (2015). *Undoing the demos: Neoliberalism's stealth revolution.* New York: Zone Books.

Butler, J. (1997). *The psychic life of power: theories in subjection.* Stanford, California: Stanford University Press.

Carter, L. (2011). Gathering in threads in the insensible global world: The wicked problem of globalization and science education. *Cultural Studies of Science Education, 6,* 1–12.

Deleuze, G. (1988). *Foucault.* University of Minnesota Press.

Foucault, M. (1982). The subject and power. In H. L. Dreyfus & P. Rabinow (Eds.), *Michel Foucault: beyond structuralism and hermeneutics* (pp. 208–226). Chicago: University of Chicago Press.

Foucault, M. (1985). *The use of pleasure.* New York: Pantheon Books.

Foucault, M. (1986). *The care of the self* (The history of sexuality, vol. 3). New York: Vintage Books.

Foucault, M., & Senellart, M. (2010). *The birth of biopolitics: Lectures at the college de France, 1978–1979.* New York: Picador.

Freire, P. (2000). *Pedagogy of the oppressed.* New York: Bloomsbury Publishing.

Harding, S. (1993). *The science question in feminism.* Ithaca: Cornell University Press.

Hardt, M., & Negri, A. (2000). *Empire.* Cambridge: Harvard University Press.

Hodson, D. (2003). Time for action: Science education for an alternative future. *International Journal of Science Education, 25,* 645–670.

Jameson, F. (1991). *Postmodernism, or, the cultural logic of late capitalism.* Durham: Duke University Press.

Lazzarato, M. (2013). Enunciation and politics. In S. O. Wallenstein & J. Nilsson (Eds.), *Foucault, biopolitics, and governmentality* (pp. 00–00). E-print: Stockholm.

Lipman, P. (2013). *The new political economy of urban education: Neoliberalism, race, and the right to the city.* Chicago: Taylor & Francis.

Rancière, J., & Corcoran, S. (2010). *Dissensus: On politics and aesthetics.* London: Continuum.

Rancière, J. (2006). The ethical turn of aesthetics and politics. *Critical Horizons, 7*(1), 1–20.

Roth, W. M., & Barton, A. C. (2004). *Rethinking scientific literacy.* Psychology Press.

Žižek, S. (2011). *Living in the end times* (Rev. pbk. ed.). London: Verso.

Afterword
Different Concepts and Tools Will Be Needed to Bring About an Ethically and Politically Engaged Science Education

Abstract I conclude this book with a brief call to subvert science education as a controlling conservative discipline. I suggest two concepts: creative, critical ontologies and the commons as tools that can steer science education in more just directions.

The main purpose of this book was to position subjectivity as a primary consideration when thinking about an ethically and politically engaged science education for twenty-first century problems. What constitutes, animates, and fosters subjectivities in education is an open sociopolitical question, yet educators need to proceed with this question in mind if they are going to contribute toward just futures. Scholars in both science and mathematics education are now calling for educators to position sociopolitical concerns as central to educational research, policy, and curriculum (Gutiérrez 2013; Tolbert and Bazzul 2016). We may colloquially call this the *sociopolitical turn* in educational research, one that builds off the cultural and linguistic turns that took hold of social science research in the 1990s—perhaps marked best in science education by Lemke's (1990) book *Talking Science*. This sociopolitical turn is less something declared by researchers, and more something immanent to our current historical moment—where more and more people recognize the vital importance of progressive, emancipatory politics to social and environmental justice. To put it in the language of Chap. 4, there are (bio)political forces at work that oppose the oppressive forces of biopower, forces that are larger than any discipline or institution of education. Educational research in both mathematics and science education can no longer remain separate from these forces; even when conservative forces from governments and research organizations wish otherwise!

For science education to properly engage twenty-first century problems, new concepts and tools must be employed in the service of ecologically and socially just futures. In this book I have already discussed several–the discursive constitution of subjectivities, dimensions of the ethical self, biopolitics as an analytical frame, and the revitalization of politics as a guiding ethical principle. And these are only a small fraction of what is possible and needed for science education to engage

© The Author(s) 2016
J. Bazzul, *Ethics and Science Education: How Subjectivity Matters*,
SpringerBriefs in Education, DOI 10.1007/978-3-319-39132-8

ethically and politically toward sustainable futures. To conclude this book, I would now like to lay out two theoretical concepts that can help guide science education research toward such futures. These are *Creative, Critical Ontologies* and *the Commons*.

Creative, Critical Ontologies

One problematic aspect of this book on ethics is its anthropocentric focus on human language, and exclusion of the ways in which non-human non-living entities and actants can relate to ethics. Critical scholars in science education have begun to turn toward (new) materialist approaches that focus on the relations between a wide variety of objects, affects, people, animals, materials, etc., as a means to critically understand our current social, ecological moment. With this shift comes a focus on difference, being, and becoming, in relation to others. A key concept that I have used is Deleuze and Guattari's (1987) notion of the *assemblage*, which involves imagining the world as a series of interconnected forces, material components, and discourses that can be reworked, created and changed (Bazzul and Kayumova 2015). Students and Teachers can map the material and discursive relations that construct our ecological and social world and look for ways to effect change. Viewing the world–as assemblage–can also change the way human beings typically see ethical action and decision making because we can better view human life as fully belonging to an integrated world where other entities become our responsibility.

Let us go a little deeper. Barad (2007) demonstrates in *Meeting the Universe Halfway* that entities should not be seen as distinct from the phenomena that produce them—more specifically that entities are simultaneously the result of *intra-actions* between entities (with)in phenomena. That is, entities do not so much inter-act as *intra-act* within a phenomenon, co-constituting each other. Seeing the world as an entangled assembly of things, where the identity, constitution and even well-being of one entity is contingent on others, means also that we are ethically entangled with everything else. In such an ontological view, human beings cannot divorce themselves from the biodiversity of the planet, as well as the sociopolitical contexts that continue to create and constitute others, be they biota or abiota. It also cautions us against thinking there is a universally 'right' and 'wrong' way to act in situations because unique phenomena, human-induced climate change for example, require specific actions for the well-being of both humans and non-humans. According to Barad, ethics in an intra-active entangled world allows entities to respond and flourish within a range of options. What could it mean to allow rivers, frogs, forests and soil to respond? This includes being accountable to both past and future; a concept that has been an integral part of indigenous ways of knowing for centuries (Mohawk 2010). What is common to Barad's (2007) notion of entanglement, and Deleuze and Guattari's assemblages, is the break down of modern dualities such as human/non-human, nature/culture, and natural/un-natural upon which so many ecological and social inequities are based (Barad 2012).

Turning to material arrangements, their nature, and what they allow us to do or become, represents a turn toward critical ontology, perhaps a long ignored theoretical consideration in scholarship related to pedagogy and curriculum. Thinking in a critical, ontological way can involve both what is actual, or here before us, and what is virtual, things that have not yet come to exist but for which we hold out hope.

The Commons

'The commons' are those aspects of social, cultural, and biological communities that are accessible by all, including collective institutions, communities, virtual spaces, parks, etc. Now that we have entered, officially or unofficially, into what many scientists are calling the *Anthropocene* (Lewis and Maslin 2015) (humans as a geological force) it is more important than ever to reconceptualize, protect and produce our shared world in-common. The commons include both the natural and social commons, what is shared by all, but also what is produced by all, for the sake of all (Hardt and Negri 2009). Parks are commonly seen as the natural commons, but they are also part of the social commons because they are produced by a multiplicity of people for people. The breakdown between what constitutes the "natural" and "social" world is key to preserving and producing what is common precisely because, in the Anthropocene, behaviour and practices deemed cultural, social, political, and historical such as colonialism, the industrial revolution, or World War II, have marked the atmosphere, the geological record, and the evolutionary history of earth in very pronounced ways.

The shared commons can provide hope against the destruction of the planet and those elite global forces that seek to harness the power of the many, through commodification of the commons, for the benefit of a few. Education can nurture a world-in-common by challenging the exploitation of the common for private interests, which includes the production of profit and private capital. Indeed, resistance to exploitation of the commons will come from below through the continued production and preservation of all that is shared in common, from scientific knowledge to communal land and water (Henderson and Hursh 2014). As Hardt and Negri put it, "only a multitude can produce the common" (p. 303); and today we see a multiplicity of peoples producing what is common through collective struggles of land/water use, human rights, racial and social equality, information sharing, as well as economic justice. Preserving and producing the commons is not just about stopping its commodification, but ensuring that the subjectivities necessary for its continual renewal are formed with these goals in mind. Instead of education reproducing the subjectivities necessary for capitalist exploitation of the natural and social commons, education must be geared towards creating and maintaining what is common and shared by all. For example, science education can provide the common knowledge(s) needed to nurture and grow the common. Knowledges must be made available to educational communities if they are to organize around the central concerns of our time: climate change; inequality; preserving biodiversity; sustainable modes of living; and the freedom for people to be, and become, different. In other words, the building of alternative futures. And make no mistake, making education work for principles such as sustainability is more than difficult in neoliberal times (Miller 2016; McKenzie et al. 2015).

Toward a Transdisciplinary, Politically Engaged Science Education

To meet the urgent ecological and social needs requires science education to transform itself by valuing diversity in thinking and insubordination in the face of conformity and forced standardization. I'd like to conclude this research brief on subjectivity and ethics with a short list of what science educators and educational scholars can do to shift the field toward more ethically and politically engaged practices.

Focus on the Sociopolitical—Make sociopolitical concerns such as racial inequality, economic exploitation, and the destruction of Earth's diversity the focus of science education pedagogy, curriculum, and research.

Recognition that our Current Methods are Inadequate—Look around. How the mainstream science education community does science education (myself included) does not provide the necessary ethical guidance to halt environmental destruction and work toward just futures. Only some ideas and methods are legitimized.

Flatten Hierarchies—The science education community, at least in the United States, is full of people who would build paradigms around their work, and the work of their friends, to the detriment of new, critical scholars. Voice must be given to marginalized scholars who have much to contribute.

Challenge Our Privilege—Academia is always already a privileged place, especially if one is white, male, and not in poverty (like I am). Academics must continually negotiate and challenge the inherent position of power we occupy by siding with the oppressed and living things under attack.

Nurture a Healthy Disrespect—Simply put, listen to the voice of indignation that tells you something is unjust with the establishment, and work with those who will join you to align science education with the goals of justice.

I'll stop the list at five. As educators we need to understand how much potential power resides in education in terms of producing the subjectivities necessary for a just world. No one educator is capable or realizing or 'figuring out' the monumental challenge of gearing education toward justice. It will require a multitude.

References

Barad, K. (2007). *Meeting the university halfway: Quantum physics and the entanglement of matters and meaning*, Durham: Duke University Press.

Barad, K. (2012). Interview. In R. Dolphijn & I. Van der Tuin (Eds.), *New materialism: Interviews and cartographies* (pp. 48-70). Ann Arbor: Open Humanities Press.

Bazzul, J., & Kayumova, S. (2015). Toward a social ontology for science education: Introducing Deleuze and Guattari's assemblages. *Educational Philosophy and Theory*, 1-16.

Gutiérrez, R. (2013). The sociopolitical turn in mathematics education. *Journal for Research in Mathematics Education*, *44*(1), 37–68.

Hardt, M., & Negri, A. (2000). *Empire*. Harvard University Press.

Hardt, M., & Negri, A. (2009). Commonwealth. Harvard.

Henderson, J. A., & Hursh, D. W. (2014). Economics and education for human flourishing: Wendell Berry and the Oikonomic alternative to neoliberalism. *Educational Studies*, *50*(2), 167–186.

Lemke, J. L. (1990). Talking science: Language, learning, and values. Norwood, NJ: Ablex.

Lewis, S., Maslin, M. (2015). Defining the Anthropocene. *Nature*, *519*, 171–180. doi:10.1038/nature14258.

McKenzie, M., Bieler, A., & McNeil, R. (2015). Education policy mobility: reimagining sustainability in neoliberal times. *Environmental Education Research*, *21*(3), 319–337.

Miller, H. K. (2016). Undergraduates in a sustainability semester: Models of social change for sustainability. *The Journal of Environmental Education*, *47*(1), 52–67.

Mohawk, J. (2010). *Thinking in Indian: a John Mohawk reader*. In J. Barreiro (Ed.). New York: Fulcrum Publishing.

Tolbert, S., Bazzul, J. (2016). Toward the sociopolitical in science education. *Cultural Studies of Science Education*. (online first).